CENTRIC 2018

The Eleventh International Conference on Advances in Human oriented and Personalized Mechanisms, Technologies, and Services

October 14 - 18, 2018

Nice, France

CENTRIC 2018 Editors

Stephan Böhm, RheinMain University of Applied Sciences - Wiesbaden, Germany

Printed from e-media with permission by:

Curran Associates, Inc.
57 Morehouse Lane
Red Hook, NY 12571

Some format issues inherent in the e-media version may also appear in this print version.

Printed by Curran Associates, Inc. (2018)

International Academy, Research, and Industry Association (IARIA)
412 Derby Way
Wilmington, DE 19810

Phone: (408) 893-6407
Fax: (408) 527-6351

petre@iaria.org

Additional copies of this publication are available from:

Curran Associates, Inc.
57 Morehouse Lane
Red Hook, NY 12571 USA
Phone: 845-758-0400
Fax: 845-758-2633
Email: curran@proceedings.com
Web: www.proceedings.com

Table of Contents

CENTRIC 2018

Forward

The Eleventh International Conference on Advances in Human-oriented and Personalized Mechanisms, Technologies, and Services (CENTRIC 2018), held on October 14 - 18, 2018- Nice, France, addressed topics on human-oriented and personalized mechanisms, technologies, and services, commonly known as I-centric.

There is a cohort of technologies that favored the so called "user-centric" services and applications. While some of them reached some maturity, others are to prove their economics (WiMax, IPTV, RFID, etc). The human-oriented and personalized technologies and services rely on a key set of features, some to be deployed, others getting more mature (personal profiles, preferences, identity, proximity, personal devices, etc.). Following, advanced applications covering human related activities benefit from personalized and human-oriented networks and services, especially preventive and personalized medicine, body networks and devices, or anticipative systems.

The conference provided a forum where researchers were able to present recent research results and new research problems and directions related to them. The conference sought contributions presenting novel result and future research in all aspects of user-centric mechanisms, technologies, and services.

Similar to the previous editions, this event continued to be very competitive in its selection process and very well perceived by the international community. As such, it attracted excellent contributions and active participation from all over the world. We were very pleased to receive a large amount of top quality contributions.

We take here the opportunity to warmly thank all the members of the CENTRIC 2018 technical program committee as well as the numerous reviewers. The creation of such a broad and high quality conference program would not have been possible without their involvement. We also kindly thank all the authors that dedicated much of their time and efforts to contribute to the CENTRIC 2018. We truly believe that thanks to all these efforts, the final conference program consists of top quality contributions.

This event could also not have been a reality without the support of many individuals, organizations and sponsors. We also gratefully thank the members of the CENTRIC 2018 organizing committee for their help in handling the logistics and for their work that is making this professional meeting a success.

We hope the CENTRIC 2018 was a successful international forum for the exchange of ideas and results between academia and industry and to promote further progress in personalization research. We also hope Nice provided a pleasant environment during the conference and everyone saved some time for exploring this beautiful city.

CENTRIC 2018 Steering Committee

Lasse Berntzen, University College of Southeast, Norway
Stephan Böhm, RheinMain University of Applied Sciences - Wiesbaden, Germany
Mu-Chun Su, National Central University, Taiwan
Florian Volk, Technische Universität Darmstadt, Germany
Nitin Agarwal, University of Arkansas at Little Rock, USA

CENTRIC 2018 Industry/Research Advisory Committee

Vitaly Klyuev, University of Aizu, Japan
Anbang Xu, IBM Research - Almaden, USA
Sabine Coquillart, INRIA, France
Thomas Marx, University of Applied Sciences Bingen, Germany

CENTRIC 2018

Committee

CENTRIC Steering Committee
Lasse Berntzen, University College of Southeast Norway, Norway
Stephan Böhm, RheinMain University of Applied Sciences - Wiesbaden, Germany
Mu-Chun Su, National Central University, Taiwan
Florian Volk, Technische Universität Darmstadt, Germany
Nitin Agarwal, University of Arkansas at Little Rock, USA

CENTRIC Industry/Research Advisory Committee
Vitaly Klyuev, University of Aizu, Japan
Anbang Xu, IBM Research - Almaden, USA
Sabine Coquillart, INRIA, France
Thomas Marx, University of Applied Sciences Bingen, Germany

CENTRIC 2018 Technical Program Committee

Nitin Agarwal, University of Arkansas at Little Rock, USA
Mohammed A. Akour, Yarmouk University, Jordan
Mehdi Ammi, University of Paris-Saclay - CNRS/LIMSI, France
Alessio Angius, Polytechnic of Milan, Italy
Angeliki Antoniou, University of Peloponnese, Greece
Aslihan Tece Bayrak, Media Design School, New Zealand
Lasse Berntzen, University College of Southeast Norway, Norway
Stephan Böhm, RheinMain University of Applied Sciences - Wiesbaden, Germany
Marco Bruti, Telecom Italia Sparkle S.p.A., Rome, Italy
Maher Chaouachi, McGill University, Canada
Daniel B.-W. Chen, Monash University, Australia
Sabine Coquillart, INRIA, France
José Luis Cuevas Ruíz, Instituto Federal de Telecomunicaciones, Mexico
Geeth De Mel, IBM Research, UK
Giorgio De Michelis, University of Milano - Bicocca, Italy
Yoanna Dosouto, Alaska Airlines, USA
Rainer Falk, Siemens AG, Germany
Fabio Gasparetti, Roma Tre University, Italy
Huan Gui, University of Illinois at Urbana-Champaign, USA
Riccardo Guidotti, KDD Lab | ISTI-CNR, Pisa, Italy
Till Halbach, Norwegian Computing Center, Norway
Qiang He, Swinburne University of Technology, Australia
Reiko Hishiyama, Waseda University, Japan
Tzung-Pei Hong, National University of Kaohsiung, Taiwan
Marius Rohde Johannessen, Universty of South-Eastern Norway, Norway
Imène Jraidi, University of Montreal, Canada

Parves Kamal, St. Cloud State University, USA
Saman Kamran, University of Lugano, Switzerland
Teemu Kanstren, VTT, Finland
Vitaly Klyuev, University of Aizu, Japan
Jaroslaw Kozlak, AGH University of Science and Technology, Poland
Daniela Marghitu, Auburn University, USA
Célia Martinie, IRIT, University Toulouse 3 Paul Sabatier, France
Erik Massarczyk, University of Applied Sciences – RheinMain - Wiesbaden Rüsselsheim, Germany
Jose Manuel Molina Lopez, Universidad Carlos III de Madrid, Spain
Thomas Marx, University of Applied Sciences Bingen, Germany
Yoshimasa Ohmoto, Kyoto University, Japan
Valentin Plenk, Hof University - Institute of Information Systems (iisys), Germany
Valentim Realinho, Polytechnic Institute of Portalegre, Portugal
Ann Reddipogu, RKode Ltd, UK
Michele Risi, University of Salerno, Italy
Elad Michael Schiller, Chalmers University of Technology, Göteborg, Sweden
Monica M. L. Sebillo, University of Salerno, Italy
Jingbo Shang, University of Illinois, Urbana-Champain, USA
Jungpil Shin, University of Aizu, Japan
Roman Shtykh, Yahoo Japan Corporation, Japan
Jacques Simonin, IMT Atlantique, France
Christophe Strobbe, Hochschule der Medien, Stuttgart, Germany
Mu-Chun Su, National Central University, Taiwan
Jörg Thomaschewski, University of Applied Sciences Emden/Leer, Germany
Yuan Tian, University of Texas at Dallas, USA
Tsuda Toshitaka, Waseda University / Fujitsu Limited, Japan
Jacques Traoré, Orange Labs, France
Carlos Travieso González, University of Las Palmas de Gran Canaria, Spain
Florian Volk, Technische Universität Darmstadt, Germany
Benjamin Weyers, RWTH Aachen University, Germany
Sanjaya Wijeratne, Kno.e.sis Center | Wright State University, USA
Anbang Xu, IBM Research - Almaden, USA
Sherali Zeadally, University of Kentucky, USA

Copyright Information

Approach to Develop an Assistant Application for Controlling Trace Accuracy in Travel Timelines

Andrei Kuznetsov
T-Systems Rus
Saint-Petersburg, 199034, Russia
E-mail: andrei.kuznetcov@t-systems.ru

Evgeny Pyshkin
University of Aizu
Aizu-Wakamatsu, 965-8580, Japan
E-mail: pyshe@u-aizu.ac.jp

Abstract—**Accurate location information collected during a trip is crucial for many post-travel activities. In the digitalized world, many of these activities (such as annotating pictures) are supported by different location-aware applications. Since these applications are also used in non-travel related scenarios, the applications cannot "know" in advance, what the appropriate location accuracy level is. In this paper, we analyze a couple of general purpose Google ecosystem applications in the context of post-travel activities with the use of real data collected during a one week trip. We examine the possible scenarios where location accuracy is not high enough to fulfill the user requirements, and propose a novel approach, which allows the location-aware application receiving more accurate location data according to policies set by the user. In our model (which is naturally applicable not only to Google ecosystem applications), we combine a general-purpose application with an assistant application aimed at managing location data quality. We describe the implementation of a prototype companion application and demonstrate that this application allows travelers to continue using regular applications (such as Google Timeline) with achieving the desired level of location accuracy.**

Keywords–Mobile application; Intelligent assistant; User behavior modeling; Location data; Accuracy; Human-centered design.

I. Introduction

Digitalization significantly affects the domain of services and tools for travelers, including web services and mobile applications. Emerging solutions are currently addressing the process of creating applications, which are not about simple time-budget optimization: the users expect to get features allowing travelers to develop and share their own emotionally intense and memorable experience [1].

Traveling is much more than simply moving from one point to another with visiting some attractions and sightseeing spots. While traveling, people are involved in numerous activities connected to a variety of ubiquitous digital services and applications available via mobile devices. At the different stages of a journey, the travelers require different kinds of assistance.

Before the trip, the "high-level plans" (what countries or cities we want to visit) meet the "low-level" arrangements (what are the particular places to be visited in a city or its district, how should we book such visits, are there visa requirements, etc.), as well as some minor (but still important) organizational matters (packing your camera, buying electric power converters, reading tourist guides, etc.).

There are many desktop, mobile and web applications for trip preparation. For example, in [2], the authors combine a recommendation algorithm with interactive route visualization for creating and managing personalized itineraries. In [3], the authors address the needs of those professional and amateur guides who are interested to be creators of the tours and excursions accessible from travelers' devices. In [4], the system for planning multi-day personalized travel tours is described. This system uses information gathered from a travel-centric social network.

In the course of a journey, an emphasis is shifted to the on-the-fly planning aspects. They include public transportation itineraries, navigation, checking opening hours, finding good places for eating, making arrangements conditioned by possible schedule changes (such as missing the train, decisions to stay longer, experiencing flight delays, unexpected business meetings, strikes, etc.). There is also an important aspect of discovering places which are worth visiting *during a desired period of time*. In [5], the authors described a trip builder application, which uses a collection of geo-tagged photos from Flickr [6] as a spatiotemporal information source for planning personalized sightseeing tours in cities. Kachkaev and Wood [7] advanced an approach to creating walking tours lasting during a desired time slot. They also used crowd-sourced photography contents with paying attention to non-trivial algorithms of photo qualitative analysis and suggested a filtering method for removing irrelevant images from the dataset with the use of anomaly detection in spatial and temporal distribution of photographs. Using geotagged photos for trip planning and location recommendation can also be found in a number of other works [8][9]. To sum up, most of the above-mentioned tools are focused on better user personalization, travel itinerary customization, as well as leveraging experience of other travelers.

There is a reasonable number of **after-journey activities** including sorting pictures, recordings, notes and other material and media artifacts that one wants to store or share.

In regards to post-travel experience, most of the traveler-oriented scenarios are about timing and location. Surprisingly, exact and accurate location data might sometimes become even more important in *after-journey* activities than during the trip. Indeed, if we have precise information about the traveler's locations in time, many other parameters (like a place name, city, country, etc.) can be calculated. Subsequently, these calculated parameters can be used to annotate pictures, notes or other records, including those, which are produced by location-unaware legacy devices, e.g., by cameras with no built-in or installed GPS (Global Positioning System) sensor.

In this paper, we specifically address such a post-travel experience, with a particular attention on the problem of collecting user location information and using it for travel-oriented use cases. The objective of this work is to describe a design pattern aimed at improving behavior of existing location-aware applications. The pattern includes a companion application, which is aware about user context (e. g., "the user is traveling") and any number of other applications. The companion takes care about location data accuracy, while the other applications are focused on their major usage scenarios (like putting pictures on the map or building photo albums for each visited place). We separate concerns of the different components. In so doing, we expect to improve user experience by assuring collecting more accurate location data when they are really needed, and conserving battery energy when they are not.

The remaining text is organized as follows. In Section II, we analyze the problem of collecting user location data and describe the state-of-the-art approaches addressing a tradeoff between accuracy and energy consumption. In Section III, we examine a sample use case of annotating pictures (taken during a trip) with location and sightseeing information. In Section IV, we introduce a sample dataset (including location data and photograph metadata for one week trip). In Section V, we demonstrate (with the help of our sample dataset) that the data collected by a general purpose application might not be enough to reconstruct tourist route with an accuracy suitable for annotating collected records (e.g., pictures). Section VI introduces a companion application, which enables general purpose location-aware applications to receive location data with higher accuracy according to policies set by the user. Section VII concludes the paper.

II. Related Work

Though the problem of collecting user location data is not new, it is still an issue for mobile software development. A tradeoff between accuracy and energy consumption is an important factor of the location data collection process: nowadays smart device users search more for power plugs than for network connections [10]. This consideration is important for electronic tourist diary collection systems as well: additional pressure on the battery (conditioned by GPS sensors) "can lead to an unacceptable battery consumption for users" [11]. There are three major approaches used to handle this issue.

The first approach is to *optimize the quality of service for given power consumption level* (aimed at getting better accuracy with no additional pressure on the battery). For example, in [12], the authors suggest using smartphone's built-in accelerometer, gyroscope and magnetometer to improve location accuracy. Calibration is implemented by using a number of distinguishable patterns (such as going up/down, stopping at traffic lights, etc.). Though the major focus of that work is on location accuracy, the authors reported that in addition to location accuracy improvement they succeeded to reduce power consumption. This became possible because of an algorithm used to disable the GPS sensor under certain conditions, while keeping tracking locations with the use of inertial sensors only.

The second approach is to *optimize power consumption for a given quality of service requirements*. This approach usually suggests the use of alternative (less power consuming) sensors for obtaining user's locations (e. g., wi-fi, mobile networks, and others). The main idea is to use an intelligent algorithm in order to select an appropriate source of location data, based on current accuracy requirements. For example, cell-towers visibility fingerprints or triangulation can be used to obtain coarse location [13]. One more way to optimize power consumption is to detect the user's current activity. Expensive location sensors can be disabled if it is known (e. g., by analyzing less expensive accelerometer data) that the device is actually not moving [14][15], or its GPS signal is blocked indoors.

The third approach is to *optimize applications and algorithms by relaxing location accuracy requirements* (e. g., by using coarse location instead of fine location whenever possible). Such an optimization normally happens on a case-by-case basis. For example, in [16], the authors describe a number of search-oriented use cases (like "find pizza stores nearby"). In order to fulfill the request, an application needs to know the user's location. Location accuracy requirements might depend on geographical density of the searched objects. For the "pizza-case", it means that in a populated area with many pizza stores, accuracy demands are more strict (100–200 meters) in contrast to far away environments with only few pizza stores (where accuracy of 1–2 kilometers is sufficient to locate the nearest shop).

In [14], the authors suggest shifting focus from the coordinates-oriented locations to place-oriented locations. In other words, the users are normally not interested in raw coordinates (longitude and latitude), but in the places they visit (such as "my office", or "5th floor cafe"). As a result, we can identify the places by using so-called fingerprint techniques (which do not only include wi-fi, Bluetooth or cellular signals, but also FM radio fingerprints [12] or ambient fingerprints such as sound, light or color [14]).

In this paper, we follow the third approach with a few important adjustments:

- Our primary goal is not to save energy, but to collect location data with an accuracy acceptable for post-travel scenarios.
- Currently we delegate the decision making about "acceptable" accuracy to the owner of the device. Our application only enforces the policies already set by the user.
- We want that **other applications** (not ours) running on the device to receive location data with an acceptable accuracy.

The objective of this contribution is to bridge the gap between the existing services and the users' needs, and, therefore, to enable the existing general-purpose services (like Google Timeline) supporting better user experience in post-travel activities. We want to give travelers an opportunity to control the captured travel route accuracy with respect to particularities of real travel-related use cases. The main challenge is not to collect data, but to share them: collected GPS coordinates should be "sharable" among other applications supporting travel use cases (including photo albums, maps with landmarks, social networks, etc.).

III. Case Study: Annotating Pictures Collected During a Trip

In this section, we present a use case of "annotating pictures" as one example of common post-travel activities. Suppose that the user is travelling with a camera and a smartphone. The

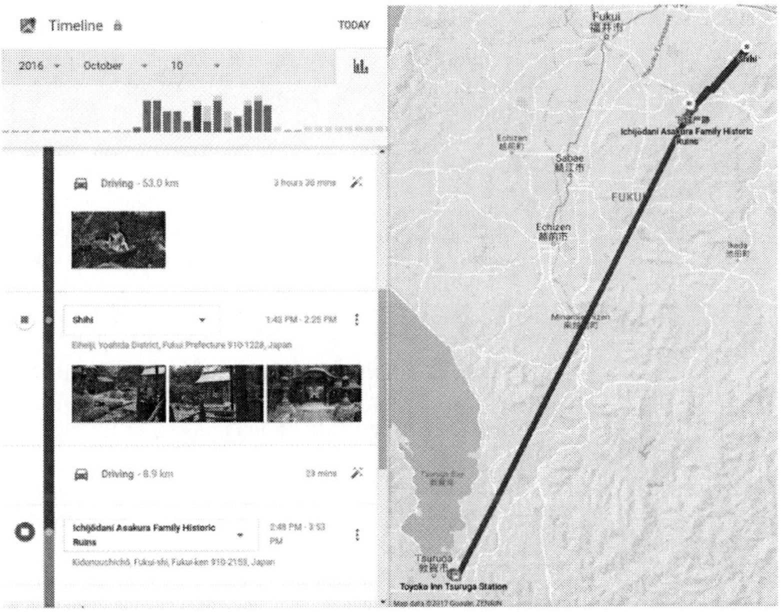

Figure 1. Google Timeline for one travel day.

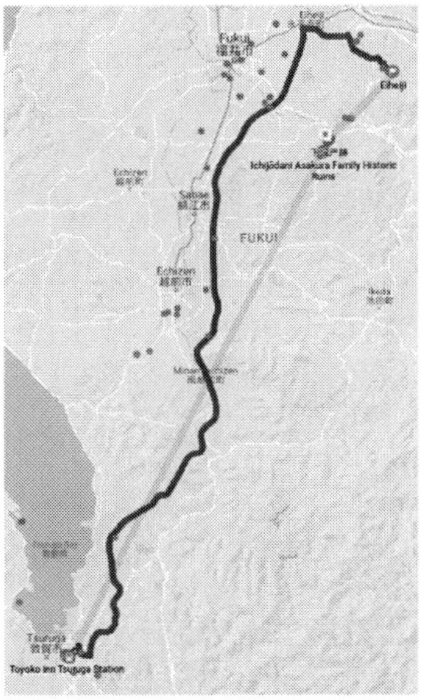

Figure 2. Google Timeline for one travel day (automatically recognized route with raw data).

latter is used for navigation and timetable information, whereas a camera is (naturally) used to take pictures of the different sightseeing places. We assume that pictures produced by the camera do not have location information, but do have timestamp information. After the user completes the journey, he or she might be interested in annotating the pictures with location metadata containing country, city and/or tourist attraction names.

To be exact, we assume that the user has an Android-based smartphone device (this is not a significant limitation nor constraint: any modern mobile operating system offers the services discussed hereafter). To proceed with attaching the corresponding place-related metadata, the user could combine Google Photos and Google Maps Timeline applications.

There are two basic options to annotate pictures with the location data:

1) Raw data can be exported from the Google Timeline in KML (Keyhole Markup Language [17]) or JSON (JavaScript Object Notation [18]) formats. Then GPS information in all the pictures can be updated using the tool `exiftool` with `-geotag` flag [19].
2) Upload all photos to the Google Photos service and let the service putting the pictures on the map. In this case, the user can see the pictures associated with a particular sightseeing place just inside the Timeline application (as Figure 1 demonstrates), or in the album of "places" in Google Photos application.

Unfortunately, due to possibly inaccurate location information, the pictures might be attributed to wrong sightseeing places. The location accuracy might be low for the reason that the Timeline application is very "conservative" in terms of battery usage. Custom GPS tracker applications would show better accuracy at a price of higher battery usage (in a short trip, such a compromise often inherently meets user expectations). The major drawback of the custom location tracking applications is that they might be not integrated well with other user's favorite applications. For example, the GPS tracker might be not integrated with a cloud photo album, therefore, the user might need finding a way to transfer location data from the tracker to the album application at his or her own (in many practical cases, no action to "transfer" location data is required, because the operating system shares location data with the other applications automatically, as we explain in Section VI).

IV. Experimental Data

In this section, we present and compare some metrics received from the Google Timeline application and a number of photographs taken during one single journey. Here is the brief summary of this journey:

- Travel dates: 10-Oct-2016 till 17-Oct-2016 (inclusive).
- Country: Japan.
- Regions: Fukushima, Fukui, Ishikawa, Kyoto, Yamanashi, and Chiba prefectures.
- Pictures total: 1112. Each picture has timestamp information and does not have any location metadata associated.
- Places visited: ≥ 13 (some places with few pictures are collapsed into bigger local areas).
- Timeline data: 3044 GPS points are collected from a smartphone. The smartphone was used to navigate in the area and to learn historical facts about the sightseeing places.
- Journey metrics:
  1) Average time interval between GPS points:
  $$\frac{8days \cdot 24\frac{hours}{day} \cdot 60\frac{min}{hour}}{3044points} < 4min/point$$
  2) Average interval of taking photographs:
  $$\frac{8days \cdot 24\frac{hours}{day} \cdot 60\frac{min}{hour}}{1112pictures} > 10min/picture$$

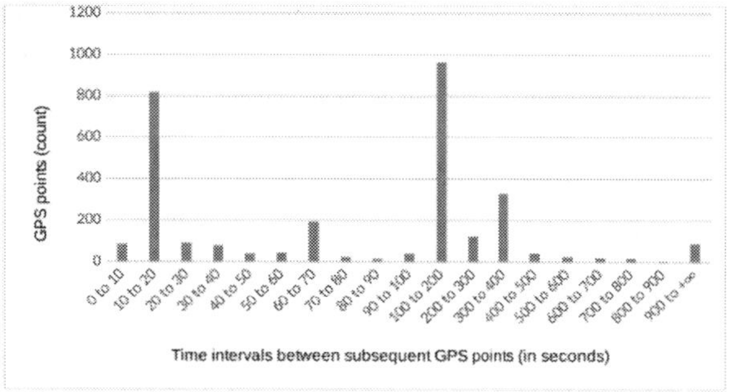

Figure 3. Distribution of timestamp intervals between subsequent GPS points recorded by the Google Timeline.

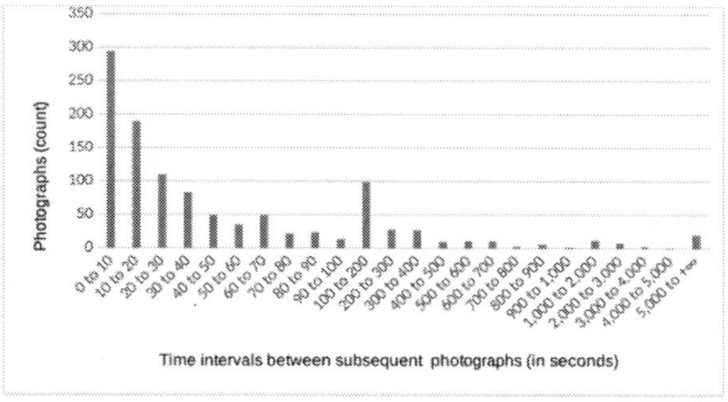

Figure 4. Distribution of timestamp intervals between subsequent photographs taken by a traveller during the trip.

As you can see, the GPS points are collected almost twice as frequent as the photographs are. One more interesting observation is on data collection frequency distribution: we sorted all the timestamps for the collected GPS points and calculated the time intervals between the subsequent timestamps. The same operation was applied to the timestamps from the collected photographs. Figures 3 and 4 show the results.

Let us note that the most pictures were taken in a series of shots made in less than 60 seconds, and there is a peak around 200 seconds. The timeline application shows nearly the same distribution: the most points are collected within the interval between 20 to 200 seconds.

Though the statistical results presented in Figures 3 and 4 seem promising, and the Timeline application is definitely very useful for post-travel experience, there is a number of interesting issues to be resolved.

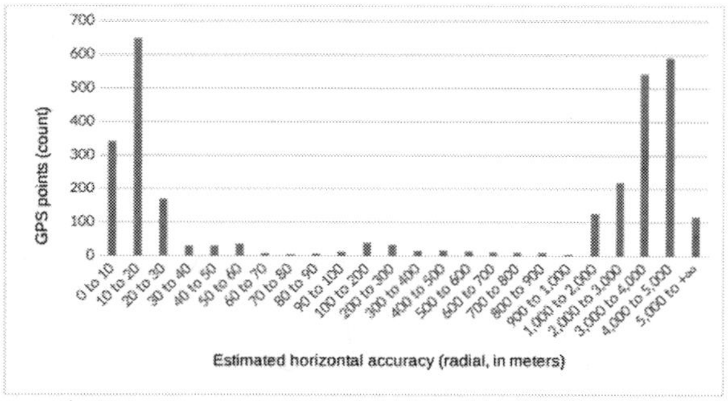

Figure 5. Distribution of accuracies of GPS points recorded by the Google Timeline.

The major issue is the significant difference between actual traveler's route (Figure 6) and track shown on the map by the Timeline application (Figure 2). We address this issue in the following section.

V. Analysis of Differences between Actual and Google Timeline Routes

Let us take a closer look at the above mentioned journey in Fukui prefecture, Japan, from Tsuruga to Eiheiji (Figure 1). By enabling "show raw data" mode, we can analyze actual points received from the location sensor (red dots in Figure 2): we can see that many points are not mapped to the route.

In principle, the Timeline application provides high-level tools that can be accessed by those who wish to fix inaccurate attribution of the visited points placed to the route (as Figure 6 illustrates).

The original (automatically recognized) route was as follows: Tsuruga–[driving]–Shihi–[driving]–Ichijodani Asakura Family Historic Ruins–[moving]–unconfirmed place–[moving]. Figure 6 shows the manually revised route. As you can see, this revised itinerary includes many places which are missing in the original track, including Fukui station, Shimokudo trace, Kasuga Shrine, Remains of Nishiyama Kousyoji temple, and Ichijodani Station.

After such a correction, the raw data (red dots in Figure 6) fit the route (blue lines in Figure 6) much better. However, it is still impossible to recognize properly, whether, for example, the Remains of Nishiyama Kousyoji temple were visited or not. In the trip dataset, we can find 5 pictures of the Remains of Nishiyama Kousyoji temple within the timeframe between 08:35:48 AM GMT to 08:37:51 AM GMT. The Google timeline contains one point within the extended timeframe between 08:33:48 AM GMT to 08:39:51 AM GMT. Nevertheless, the timeline raw data (see Figure 6) do not show that the traveler visited the temple. This issue can be explained if we analyze the raw timeline values:

```
{ "time": 8:34:25,
  "longitudeE7": 1363465709,
  "latitudeE7": 360585156,
  "accuracy": 3400 }
```

The location `accuracy` is an estimated horizontal accuracy of the location (radial, in meters), and its value is 3.4km (around 40 minutes walking distance). Since the points of interest are located close to each other (it is about only 200m from Ichijodani station to Nishiyama Kousyoji temple, and 500m to Kasuga shrine), the accuracy less than 200m is not much helpful for reliable identification of the visited places.

Figure 5 shows the overall distribution of the Timeline accuracy (during the whole trip). There are two well-distinguishable peaks: the first one is within the range $[0m, 30m]$ (about 38% of all the points fall into this range), and the second one is within the range $[1km, +\infty)$ (about 52% of all the points).

VI. Google Timeline Companion Application

Although Android is an open-source operating system, the Google Services are not. Hence, we do not know the Google Maps and Timeline internal implementation details and the algorithms they use. We expect that they are based on Google Play Services API (application programming interface) [20] for

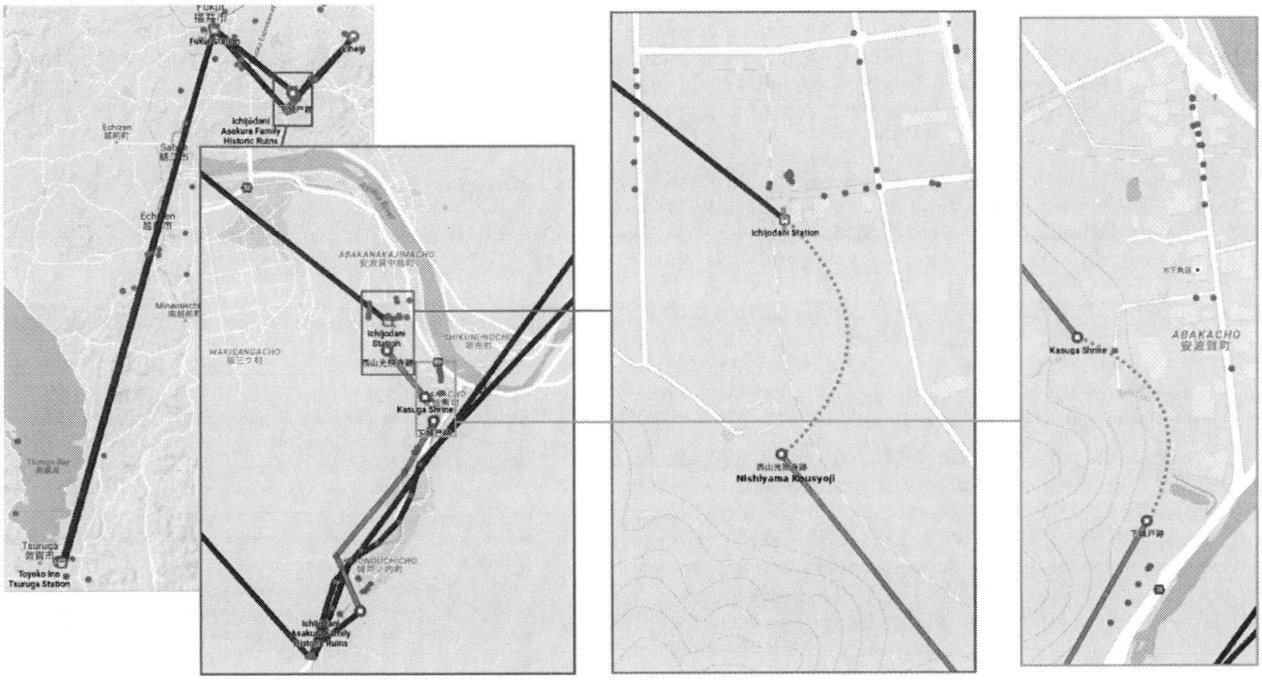

Figure 6. Google Timeline for one travel day (manually fixed route).

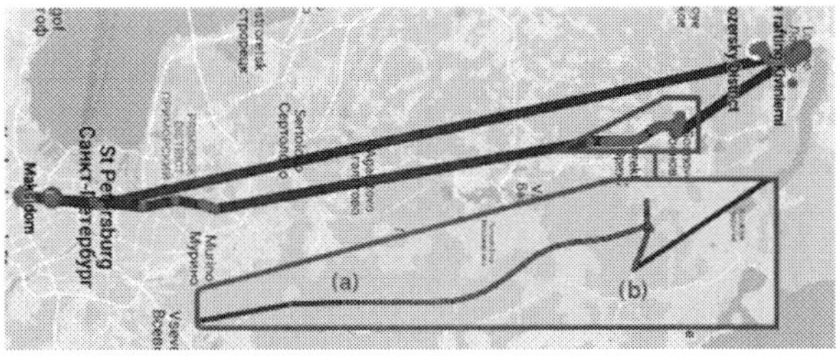

Figure 7. Side effects from turning navigation application on and off: point (a) – turned on; point (b) – turned off.

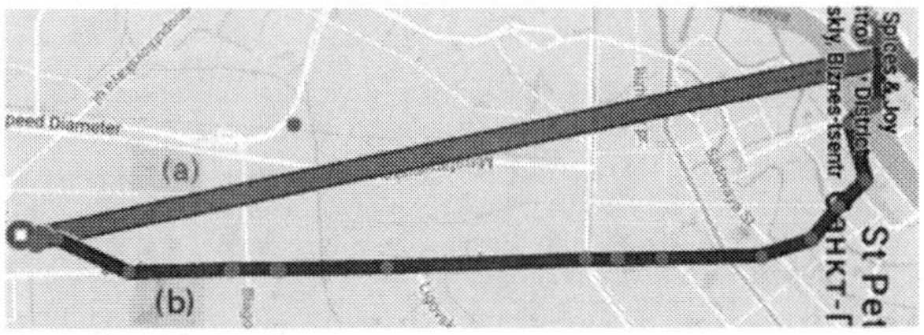

Figure 8. 40-minutes city trip (driving): (a) standard accuracy of the Google Timeline and (b) increased accuracy enabled by the companion application.

location data processing. With respect to positioning accuracy, this API provides capabilities allowing developers to proceed with fine-grained control over location accuracy, in order to save the device battery. Our assumption is that the Timeline application subscribes for all the location events if they are available (it does not activate any additional sensors), while seldom requiring coarse location and even rarely requiring fine location, which activates the GPS sensor. To be specific, in Google Services API this mode is handled via combination of `priority`, `interval` and `fastestInterval` properties of `LocationRequest` object [21]. In other words, if one application activates the GPS sensor, the Timeline application becomes getting aware of the user's location as well. Figure 7 illustrates this issue: the segment (a) to (b) corresponds to the case when a separate navigation application (Yandex.Navigator, in our case) is turned on, while the track outside this segment corresponds to the case when a separate navigation application is turned off.

Our idea is to develop a companion application, which enables the Timeline application to collect user location data more accurately. This assisting application is aware of the context: it "knows" that the user is traveling, not simply moving from one place to another. Therefore, we can subscribe for location events with an appropriate accuracy.

What makes the companion specific is that it is not aimed at receiving location information only, but it makes this information available to other location-aware applications running on the device. In fact, the companion may even ignore the received location updates without additional processing or storage, because these location events are handled by the Timeline application. The Timeline application stores location information locally, and transfers it to the cloud when the device gets online. In fact, such a companion application might not have any user interface, since it works as a service mostly. Figure 8 provides a hint on how the map timeline accuracy can be increased. Figure 8 (a) shows a Timeline track captured with no companion application running, while Figure 8 (b) shows a track captured by the Timeline application when the companion is configured to receive location updates at a higher rate (every 2 minutes).

VII. Conclusions

In this paper, we examine a use case when an (assisting) application is developed in order to improve a number of quality properties of another application. Particularly, we implemented a model of the combined usage of Google Timeline application and our companion application aimed at assuring a required level of location accuracy. It is important to note that these two applications do not interact directly with each other, but one application (Google Timeline) benefits from the fact that another one (the companion) is running.

Despite our software implementation targets the Android platform, the approach is not limited to only one platform. In present-day operating systems, the applications do not connect to any sensor directly, they use services, provided by the operating system, which shares the same data (received from a certain physical sensor) among different applications. Often (not only in Android, but also in Windows, iOS, and some others) a location service is an abstract service, which receives information from the different sources and provides a "user location", not simply "GPS coordinates".

Because of the nature of location data, there are applications that "want" to receive location information only if such information is available with no additional pressure on battery and central processing unit (CPU) usage. For example, Google Now, Siri, Cortana, and social networking applications produce outputs based on the user locations. Since these applications are popular, operating system API (anyway) supports such kind of subscribers in order to provide a balanced policy with respect to both power saving and leveraging the desired user experience.

Among the problems to be addressed in the future, we can mention a problem of finding a tradeoff between location data accuracy and battery life automatically: the location accuracy requirements might depend on the density of tourist attractions in the particular area, on the user speed, as well as on many other factors. In our current prototype, the only option to manage this issue is to configure a companion manually by using the configuration user interface we provided.

ACKNOWLEDGMENT

This work is supported by the grant 17K00509 of Japan Society for the Promotion of Science (JSPS) "Advancing interfaces, ontologies and algorithms for traveler-centric information systems supporting geographical, cultural and historical perspectives".

REFERENCES

[1] S. Park and C. A. Santos, "Exploring the tourist experience: A sequential approach," Journal of Travel Research, vol. 56, no. 1, 2017, pp. 16–27.

[2] A. Yahi, A. Chassang, L. Raynaud, H. Duthil, and D. H. P. Chau, "Aurigo: an interactive tour planner for personalized itineraries," in Proceedings of the 20th International Conference on Intelligent User Interfaces. ACM, 2015, pp. 275–285.

[3] E. Pyshkin, A. Baratynskiy, A. Chisler, and B. Skripal, "Information management for travelers: Towards better route and leisure suggestion," in Computer Science and Information Systems (FedCSIS), 2016 Federated Conference on. IEEE, 2016, pp. 429–438.

[4] I. Cenamor, T. de la Rosa, S. Núñez, and D. Borrajo, "Planning for tourism routes using social networks," Expert Systems with Applications, vol. 69, 2017, pp. 1–9.

[5] I. R. Brilhante, J. A. F. de Macêdo, F. M. Nardini, R. Perego, and C. Renso, "Tripbuilder: A tool for recommending sightseeing tours." in ECIR. Springer, 2014, pp. 771–774.

[6] SmugMug Inc. Flickr. Accessed: 2018-08-17. [Online]. Available: http://www.flickr.com/

[7] A. Kachkaev and J. Wood, "Automated planning of leisure walks based on crowd-sourced photographic content," 2014, accessed: 2018-08-17. [Online]. Available: http://openaccess.city.ac.uk/4943/

[8] A. Majid, L. Chen, H. T. Mirza, I. Hussain, and G. Chen, "A system for mining interesting tourist locations and travel sequences from public geo-tagged photos," Data & Knowledge Engineering, vol. 95, 2015, pp. 66–86.

[9] O. P. Patri, K. Singh, P. Szekely, A. V. Panangadan, and V. K. Prasanna, "Personalized trip planning by integrating multimodal user-generated content," in 2015 IEEE International Conference on Semantic Computing (ICSC). IEEE, 2015, pp. 381–388.

[10] A. T. Lo'ai, A. Basalamah, R. Mehmood, and H. Tawalbeh, "Greener and smarter phones for future cities: Characterizing the impact of gps signal strength on power consumption," IEEE Access, vol. 4, 2016, pp. 858–868.

[11] A. C. Prelipcean and T. Yamamoto, "Workshop synthesis: New developments in travel diary collectionsystems based on smartphones and gps receivers," in 11th International conference on Transport Survey Methods, 2018, pp. 340–349.

[12] C. Bo, T. Jung, X. Mao, X.-Y. Li, and Y. Wang, "Smartloc: sensing landmarks silently for smartphone-based metropolitan localization," EURASIP Journal on Wireless Communications and Networking, vol. 2016, no. 1, 2016, p. 111.

[13] A. LaMarca et al., "Place lab: Device positioning using radio beacons in the wild," Pervasive computing, 2005, pp. 301–306.

[14] D. H. Kim, Y. Kim, D. Estrin, and M. B. Srivastava, "Sensloc: sensing everyday places and paths using less energy," in Proceedings of the 8th ACM Conference on Embedded Networked Sensor Systems. ACM, 2010, pp. 43–56.

[15] C.-F. Liao, Y. Fan, G. Adomavicius, and J. Wolfson, "Battery-efficient location change detection for smartphone-based travel data collection: A wi-fi fingerprint approach," in Transportation Research Board 95th Annual Meeting, no. 16-2090, 2016.

[16] K. Lin, A. Kansal, D. Lymberopoulos, and F. Zhao, "Energy-accuracy trade-off for continuous mobile device location," in Proceedings of the 8th international conference on Mobile systems, applications, and services. ACM, 2010, pp. 285–298.

[17] Google LLC. Keyhole markup language. Accessed: 2018-08-17. [Online]. Available: https://developers.google.com/kml/

[18] "The JSON Data Interchange Format," ECMA, Tech. Rep. Standard ECMA-404 1st Edition / October 2013, Oct. 2013. [Online]. Available: http://www.ecma-international.org/publications/files/ECMA-ST/ECMA-404.pdf

[19] P. Harvey. ExifTool by Phil Harvey. Accessed: 2018-08-17. [Online]. Available: https://www.sno.phy.queensu.ca/~phil/exiftool/

[20] Google LLC. Overview of Google Play Services. Accessed: 2018-08-17. [Online]. Available: https://developers.google.com/android/guides/overview

[21] ——. Google APIs for Android: LocationRequest. Accessed: 2018-08-17. [Online]. Available: https://developers.google.com/android/reference/com/google/android/gms/location/LocationRequest

The Evolution of the Customer-Centric Helpdesk: Two Case Studies

Maryam Rezaeian
University of Gloucestershire
Library, Technology and Information Services
Cheltenham, UK
Email: MRezaeian@ glos.ac.uk

Martin Wynn
University of Gloucestershire
School of Business and Technology
Cheltenham, UK
Email: MWynn@glos.ac.uk

Abstract – **This article examines how the helpdesk function has changed over the past three decades through examination of two case studies. In 1985, the European shipment of PC workstations overtook shipments of simple terminals (i.e., video display units and keyboards, with very little processing power), instigating a revolution in end-user computing, with computer users taking advantage of new word-processor, spreadsheet, graphics, email and database applications. This article looks at two snapshots of end-user computing and helpdesk operations separated by a 30 year period – one at Glaxo Pharmaceuticals in 1988, and the other at the University of Gloucestershire in 2018. This case study research finds that whilst the range of technologies requiring support has increased markedly, this has been counter-balanced somewhat by the emergence of standards and dominant products in many technology categories. It also finds that the concept of support and the role of the end-user have evolved significantly in a rapidly changing technology landscape.**

Keywords – End-user computing; helpdesk functions; customer-centricity; office systems; personal productivity tools; IT support strategy; Service Desk.

I. INTRODUCTION

The technology environment in major organisations has changed significantly in the past 30 years. In 1988, there was virtually no use of the Internet, portable computers were in their infancy, and there were no mobile phones or tablets. Although the personal computer (PC) had broken through to become the main desktop device in the more advanced organisations, local area networks were just being introduced and MSDOS was the main PC operating system in the pre-Windows age. Many of the mainstream corporate systems were bespoke (often in 3G languages like COBOL), and the main packaged software products like the SAP and Oracle Enterprise Resource Planning (ERP) systems were just starting to be taken up by the bigger corporations.

Technology support is a key issue in nearly all organisations today, and this article examines the origins and evolution of end-user computing and helpdesk support functions over the past three decades. In so doing, the article highlights the key support issues that organisations must address today. It features two case studies. First, Glaxo Pharmaceuticals, which was an advanced technology user and was seen as a leader in its rapid uptake of PC applications in the 1980s [1]. The second case study concerns the University of Gloucestershire (UoG) in 2018. This introductory section is followed in Section 2 by a brief discussion of the background to this research and the case study methodology, and sets two research questions. Sections 3 and 4 focus on the two case studies and section 5 makes some final comparisons and addresses the two research questions.

II. BACKGROUND & RESEARCH METHOD

IT services are key in ensuring the efficiency and agility of business processes [2], and within this context, the importance of a successful helpdesk in supporting corporate performance is generally accepted. Sood [3] recently noted that "the cross-functional nature of its operation means the help desk directly impacts productivity and is an essential part of what enables an agency to meet its stakeholder needs".

As early as 1992, Bridge and Dearden [4] noted "the quality of helpdesk operations can be improved by the provision of knowledge to front line helpdesk operators" and that "this could only be done effectively if AI technology is used". This early study of helpdesk operations proved prophetic as helpdesks have evolved to meet the changing demands of end-users and have used increasingly sophisticated support systems. Existing literature also highlights the increase in the range of technologies that helpdesks have to support. Gonzalez, Giachetti and Ramirez [5], for example, note that the average number of information technologies supported by central support functions has increased from 25 to 2000 in the current millennium.

This article looks at two case studies of helpdesk operations and end-use computing requirements, spanning a 30 year time gap. The case study is a widely used methodology within business research. Bryman and Bell [6] argue that the case study is particularly appropriate to be used in combination with a qualitative research method, allowing detailed and intensive research activity, usually in combination with an inductive approach as regards the relationship between theory and research. Saunders, Lewis and Thornhill [7] argue that case studies are of particular value for explanatory or exploratory investigation.

Data collection to date has been achieved through participant observation and action research. One of the authors worked at the first case study company (Glaxo Pharmaceuticals) as IT Trainer and then End-User Computing manager in the 1984-88 period. Some of the observations included here were discussed in research publications at the time [1] [8], and these have been used as secondary sources of material. The other author has worked on the IT Service Desk at the second organization (UoG) and thus has first-hand experience of the technologies deployed and the Service Desk operations. There are thus multiple sources of evidence, which as Yin [9] suggests, is one way of increasing the construct validity of case studies. At UoG, this includes participant observation and a number of internal reports [10].

Within this context, this article addresses two research questions (RQs):

RQ1. How have the support requirements of helpdesks evolved over the past 30 years?

RQ2. How has the helpdesk developed in response to these changes in the technology landscape?

III. Case Study 1: Glaxo Pharmaceuticals 1988

Overview: Between 1984 and 1988, Glaxo Pharmaceuticals saw a rapid increase in the use of PCs, which radically changed the nature of computing within the company. In excess of 1300 PCs were installed in the company's four sites at Greenford (London), Barnard Castle (Co Durham), Ware (Hertfordshire) and Speke (near Liverpool). This expansion reflected the dramatic growth and improvement in PC-based office systems during this period, which changed the nature of standard office computing functions in the company. However, in 1984, office systems were clearly a function of the HP3000 mini-computers, there being over a thousand users of these office systems in Glaxo, over 600 of which were electronic mail users. There were just a few PC-based users of spreadsheets in the sales, marketing and market research areas. By 1988, one in four staff had a PC, and of these, six out of ten had a spreadsheet, four out of ten had a graphics package and a word-processor, and three out of ten had a database package. The use of mini-computer graphics modelling and word-processing had virtually disappeared, but electronic mail remained a function of the Hewlett Packard mini-computers, there being over 2,500 users, a fourfold expansion since 1984.

Word-processing and desktop publishing: In the period 1984-88, word-processing experienced several phases of growth. In the two years after 1984, the company standardized on one main word processing system (HPWord), based on an HP mainframe or mini-computer for all secretarial/office staff. Then, in 1987-88 as the PC became the standard desktop machine rather than the terminal, users were transferred to a PC-based version (PCWord) of the mainframe package, thus minimizing the need for retraining. Then in 1988, the company embarked on a further change that would see the introduction of a more sophisticated word-processor as the standard for secretarial use. This was in part driven by the well-publicised benefits of using the so called "desktop publishing" (DTP) packages, which required a skill level normally beyond that of the average secretary, and which also required specialist workstations (an 80386 chip, and a PostScript-compatible printer) if acceptable performance was to be achieved.

This resulted in the introduction of only two desktop publishing workstations (running PageMaker and/or Ventura software packages). However, it was expected at the time that the standard document processing software available to secretaries would come to include some DTP functions such as graphics and scanned image importation, and this is indeed what happened. It was suggested that a move to the type of mid-range product in the word-processing to DTP spectrum, such as the Lotus Manuscript or Advancewrite Plus software packages was likely. The coming of Windows as the standard operating system and the gradual dominance of the Microsoft office products was not envisaged at that time.

Database and spreadsheets: Databases are possibly the most powerful end-user tools of all the functional "off- the-shelf" packages, while spreadsheets are the most commonly used. A PC survey carried out at Glaxo in May 1988 found that for every PC database system written by the company's Information Management Division (IMD), end-users had developed three systems for themselves. The PC systems developed by IMD at the request of end-users is shown in Table I. Authorisation for these systems was done on an *ad hoc* basis, and approval for resource allocation from higher management levels was not required.

A number of different spreadsheet packages had been tried by end-users, but Lotus 1-2-3 was the most commonly used.

TABLE I. End-User Computing Systems at Glaxo 1988

End-user system name	Software
Electronic faces folder	DB3+/Tencore
Medical records	DataEase
Unpublished journals	DataEase
Label reconciliation	DataEase
Materials requisition	RBase 5000
Medical terms dictionary	Custom-made with Pascal
Accident records	DataEase
Project engineer management	DataEase
Media scheduling	DataEase
Planning & budgeting	DataEase
Action reporting	DataEase

Graphics packages: Graphics packages were not as common as word-processors, but the two were increasingly used in unison as standard secretarial software. They were used mainly for departmental reports and presentations. The data was still input manually for the most part, but electronic transfer into graphics packages was on the increase as integration with mainframe databases and other office systems improved. This was to be a forerunner of the wider integration and consolidation of office productivity tools that occurred in the Microsoft era. By 1988, the main graphics package used was Freelance or Freelance Plus, which was then from the provider of the Lotus 1-2-3 spreadsheet, ensuring ease of data transfer between the two packages.

Electronic presentations systems and presentation design software: This was a significant end-user computing activity. There were a range of software packages available for electronic systems, including PictureIt and Freelance Plus, running on the so-called "IBM compatible" PCs. PictureIt enabled the user to design bar, pie, line, organization and word charts in a range of pre-determined formats. It was extremely easy to use and yet contained sufficient variety to facilitate the design of a reasonable presentation. This was particularly useful for senior management and the sales and marketing functions.

For more specialised needs, Freelance Plus was used. This was a freeform drawing package, with a range of icon libraries that could be combined with PictureIt images. Graphs could also be imported from other software packages (including Lotus 1-2-3 and Lotus Symphony). Standard 80 column/25 line text screens could also be converted to VideoShow format and edited using VIP.

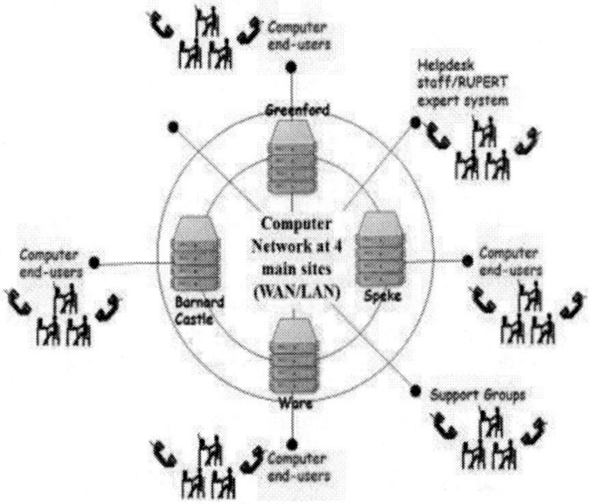

Figure 1. The Rupert helpdesk system: interaction with end-users and support groups.

The VideoShow presentation system was made available to be taken out on loan from the IMD, and each of the four sites had at least one of these machines. Having prepared the presentation with software running on the PC, this was then saved to "floppy disc" and run on the VideoShow presentation system. These presentations could be given to a large audience via a projector (e.g., Barco Data 3 or Electrohome ECP 2000) or a colour monitor for smaller audiences. The wide range of colours available (1,000) as well as the range of formats available made this a convenient way to present material suitable for a 35mm slide presentation. The obvious advantages included the portability of the presentation (one floppy disc could hold as many as 200 images) and the fact that the presentation was always in the correct order, the right way round and there were no focusing problems.

Computer based training (CBT) packages: From 1985 onwards, approximately 30 CBT packages were developed by IMD using the Tencore authoring language [11]. Most of these were for sales and marketing training, and their support and on-going enhancement and update constituted an element of PC systems support at the time.

The helpdesk function: The helpdesk function was centralized at the company's Greenford site, but had links to support staff in the company's three main manufacturing plants at Ware, Barnard Castle and Speke (Figure 1). By 1988, IMD had developed its own in-house fault logging diagnostic system, built using an expert system shell (CASSANDRA). This system was known as "Rupert" (Resolves Users' Problems Expertly).

The helpdesk had hitherto been manned by a senior network analyst who used his expertise to help solve users' problems. Rupert encapsulated some of the experts' knowledge and was able to apply it to users' problems. By asking a series of questions, Rupert could home in on a problem. In some situations it could take action such as aborting a users' session, disconnecting a terminal or asking the user to perform some action such as pressing a key etc. In other cases, where Rupert was unable to provide a full solution, the call was passed on to the support group which, in Rupert's judgement, would be best placed to deal with the problem.

The support groups were still in the main geared to helping users of the company's wide range of bespoke transaction processing and reporting systems for their manufacturing and financial functions. These were mainly written in COBOL or PASCAL, and the analyst-programmers of the day doubled up as support staff to help end-users. Indeed, for the main manufacturing system (known as "MENTOR"), there was a programme of courses run on the four sites on test machines on which the main manufacturing systems could be simulated. There were three main support groups for the main corporate business systems and a fourth for office systems and end-user computing. The main business systems were run on Hewlett Packard mini-computers at the four sites linked by a wide area network, and there were also a number of test and development machines.

In addition to helping to solve problems more quickly, Rupert also produced fault statistics which helped IMD to identify problem areas and thus continue to improve the service given to users.

In the first two months of Rupert's operation, the helpdesk staff:

- handled nearly 2,500 enquiries;
- solved 70 per cent of all queries themselves;

- found that a number of queries were from users who did not understand the applications (more training courses were arranged);
- discovered that the maintenance support provided by the terminal supplier was unsatisfactory (the company asked its supplier to modify its support system).

The major benefits of Rupert to the company were:

- its role as a training aid for new helpdesk staff;
- the ease with which new knowledge could be added to the system;
- the time taken to resolve user problems was halved;
- the improved image of IMD in the rest of the company;
- the better statistics it provided about user problems.

The last two benefits could probably have been obtained from any helpdesk function and fault reporting software. However, Rupert's excellent user interface made this a very successful application of expert system techniques. It was envisaged that the system would eventually be the focal point of a comprehensive network management system.

IV. CASE STUDY 2: UNIVERSITY OFGLOUCESTERSHIRE 2018

Overview: UoG is located across six sites within Cheltenham and Gloucester with 20 professional departments. The Library, Technology and Information Service (LTI) department provides supports for both staff and students, particularly for teaching and learning, along with the provision of appropriate training and skills development. The University has over 1,500 staff, most of which are computer users, and approximately 10,000 students, who use a range of applications on University equipment in labs and classroom environments. The IT Service Desk is located within LTI and provides full support for staff in University hardware, communications and software solutions. Support for students encompasses Office 365, assignment submission, the Moodle learning management system, and a range of IT guides accessible via MyGlos Help (a web portal guidance page which helps student to search for guidance and information).

Office productivity tools and end-user computing: Microsoft Office 2016, Adobe, SSRS, SPSS, and NVivo are the main packages that are increasingly used as standard on a daily basis. SSRS (SQL reporting) is mainly used for departmental reports, whilst SPSS and NVivo are only used for teaching and research purposes, and PowerPoint (part of Office 2016) is the main package used for presentations. There are many different packages on different machines, depending on department needs. For example, there are 150 graphics package users in the departments of art and design and landscape architecture. The operating system for the PCs is currently Windows 7, although a University-wide upgrade to Windows 10 is currently being rolled out. The University email system is based on Microsoft Office365 and hosted externally. The University supports Office apps such as Skype for business, Outlook, OneDrive, and uses the international roaming service called Eduroam to provide WiFi connectivity. Eduroam allows UoG users to login at any participating institution using their UoG login name and password. Eduroam also allows users from any participating institution to login to UoG using their local login name and password. LTI use Gmetrix to provide Microsoft office training to both staff and students. UoG supports staff and student research efforts with SPSS and Nvivo.

As regards telephony, the internal telephone system (an Avaya IP phone system providing telephony for all the University campuses and the majority of the student halls of residence) is complemented by a number of exchange lines direct from the BT exchange, which are used for alarm lines, swipe machines for debit and credit cards, and payphones around the University and in halls of residence. LTI are responsible for managing mobile phone services, which are coordinated through a centralised agreement with Vodafone. The University will provide support for equipment and software which is procured by the University, but does not support mobile phones, tablets or other equipment purchased by staff or students themselves. Nevertheless, the frontline support teams will endeavour to help students with their own devices if they can (e.g., to reinstall software or attempt data recovery), but they will not attempt to fix any major hardware or mechanical problems.

UoG Main Business Systems: There are about 60 business systems running across the University, including Sunrise, SITS Student Records, Resource link, Agresso Finance and Moodle (Table 2). All of these are now supported by LTI, although some started as departmental end-user systems prior to the centralisation of IT support within the University and the imposition of certain policies and standards. Many of these systems are administered by end-users who undertake data maintenance and general support tasks.

TABLE II: MAIN BUSINESS SYSTEMS SUPPORTED BY UOG LTI

System	Description
Sunrise	IT application to manage enquiries from students and staff
SITS Student Records	SITS is a student records management system used to store, administer and manage all aspects of student information from initial enquiry and application through to degree congregation. A configurable package from software provider Tribal.
ResourceLink	Resourcelink is an integrated HR and Payroll software package.
Agresso Finance (Live)	A global accounting system from software provider Unit4.
Moodle	Moodle is a free and open-source learning management system written in PHP and distributed under the GNU General Public License.

The SITS student records management package is one of the University's core systems, and the system is upgraded regularly with modifications and new releases from the software supplier (Tribal). These are tested and implemented in the test environment by the SITS users. When the software has been tested thoroughly and approved, a change control is raised which then goes to a change control board, who will approve or reject the change. New developments are driven by the University's business and legal requirements.

The general policy for the procurement of new software applications is that it should be based on web enabling technologies that will assist in the development of a University-wide Managed Learning Environment (MLE). This principle guides procurement when the University has the opportunity to replace systems software through the annual IT capital programme.

UoG IT Service Desk function: Sunrise is the main system used by the LTI staff to manage enquiries from students and staff. In addition, any enquiries received via the MyGlos Help Portal are redirected to the appropriate team. Different versions of Sunrise have been used by the University since the year 2000, but all with the same backend. With the latest version of this system, keywords can be used to select the problem categories and automates the assignation to appropriate support personnel.

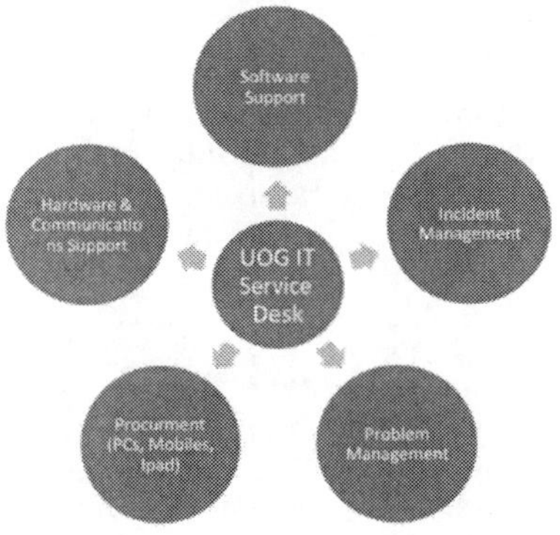

Figure 2. Main Functions of the UoG IT Service Desk 2018.

The IT Service Desk performs a number of functions (Figure 2). It has the responsibility for all user account management as well as giving access to all University business systems such as Agresso and SITS. LTI is responsible for providing the basic "image" (i.e.software footprint) for all staff and student devices. A minimum of between 4-6 weeks is allowed to enable a thorough evaluation and testing of any new software application.

When the IT Service Desk was first established in 2000, one of the front-line technicians would be on Rota to cover the helpdesk shift during the working day. Since June 2017, the IT Service Desk has been run by two officers and one manager from 8am till 5pm during weekdays. An out-of-hours service is provided by IT technicians who share the shifts across the Universitiy's three main campuses on a rota basis.

A Service Desk officer takes a call or email and logs a call. A first fix is applied if possible and remote connections used where appropriate.

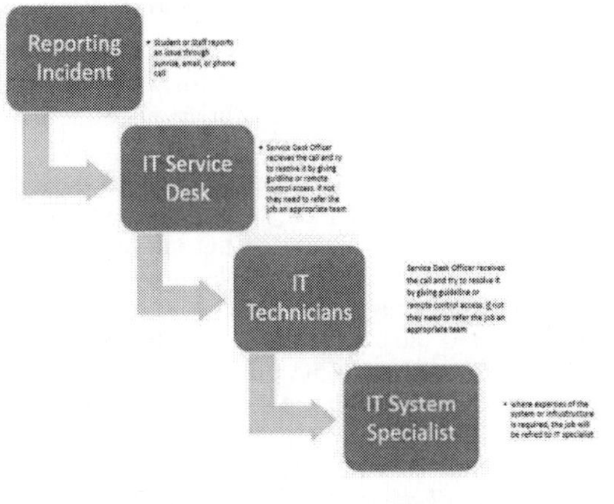

Figure 3. Incident Tracking by UoG IT Sevice Desk 2018

Figure 3 shows the escalation of a call through different levels of service expertise depending on the complexity/specialism of the problem reported. This systematic approach to tackling problems, combined with the application of dedicated human resources to solving Service Desk enquiries, has contributed to a significant improvement in response times and a more efficient IT service for the University's staff and students.

Teams across four sites use IT Service Desk tools and the Sunrise support system. The IT Service Desk tools are an integral part of the Sunrise system, and were developed as a bespoke, standalone system for UoG. Some of its main functions are:

- Password reset
- Unlock accounts
- Create guest login for externals
- Provide access to shared drives
- Deploy software
- Change voicemail passwords

LTI uses the Sunrise system to log calls, update the call, and transfer the call to the appropriate support team. Service Desk officers have access to all communications across the University by searching for the Incident Number, call details, surname, forename, category, hub area, open date, network logon, global summary, priority, escalation level, assigned group, and first time fix. The call needs to be logged under the name of the person that reported the enquiry, which can be logged by network logon (staff number) or forename and surname. The category is selected based on the enquiry; for example, if someone reports an issue with email, Service Desk officers can search for emails and pick the correct category. The use of keywords and categories ensures that an enquiry is managed by the most appropriate team. Once the category has been selected, the system will automatically pick the first line team and referral team appropriate to the job.

V. Conclusion

This section draws on the case study material discussed above to address the research questions set out in section 2.

RQ1. How have the support requirements of helpdesks evolved over the past 30 years?

Thirty years ago, the needs for IT support in major organisations were somewhat different from those of today. There was no significant use of the internet and very few mobile phones or laptops. There was no Windows - MSDOS was the main operating system for PCs. There was no SKYPE, no viruses and no wifi, but Intel chip-based PCs had established themselves in most organisations and hard-wired LANS linked them to server PCs and mini-computers. Most business systems were bespoked in-house – the age of integrated packaged software was just around the corner.

However, despite the expansion in the range of technologies that helpdesks are now called upon to support, there was arguably more variety in the range of products that needed supporting in each technology category. For example, the Glaxo helpdesk supported 5 different word-processors and several spreadsheets and graphics packages. The market was still evolving with many competing products and no obvious standards. Presentation graphics systems and videoconferencing also needed support, along with bespoked computer-based training packages in the era before on-line help functions for many software products.

There are now a greater range of technologies to support, but there are clearer standards and more obvious choices within each category. It is thus critical that the central support function has clear policies and makes product choices in each technology area. At Glaxo, despite the lack of standards in end-user software, the IMD Director was adamant that only Intel chip based PCs would be permitted in the company, and this has parallels with UoG's non-support of devices not obtained through the University procurement system. In recent years, there has been a clear imposition of standards and product choices at the University as central IT strategy and policies have taken precedence over departmental initiatives.

RQ2. How has the helpdesk developed in response to these changes in the technology landscape?

Over the time duration between the two case studies, the helpdesk function has evolved and adapted to changing requirements and developments in technology. The concept of support has also evolved, with helpdesks increasingly seeing computer users as "customers", but at the same time end-users taking some responsibility for systems ownership, data maintenance and training. The super-user and data maintenance specialists have emerged as key link personnel between the computer user-community and central IT support. The helpdesk is also increasingly seen as part of a broader Service Desk function, with service being defined as "an approach to IT service management that emphasizes the importance of coordination and control across the various functions, processes and systems necessary to manage the full lifecycle of IT services" [12]. This definition is often applied in the context of a third party service provider, but is also relevant to in-house IT service provision. At UoG, the Service Desk is the customer facing front end of all IT services which are measured against stipulated service level targets defined in service level agreements. This aligns with the IT Infrastructure Library (ITIL) concept and definition of the IT Service Desk as the single point of contact between the IT function and users, which manages incidents and service requests, and handles communication with users. There is also a more subtle change in that the service is seen as supporting business processes and people capabilities along with the pure technology elements. The Service Desk now focuses on delivering high quality customer service to end-users, whereas the helpdesk was more concerned with incident management and resolving problems related to IT in the organization [5].

The range of different technologies supported by the Service Desk has seen developments in its own support technology. In addition, the requirements set out in service level agreements have meant that Service Desks need to increase end-user satisfaction levels by responding to the incidents and problems within stipulated response times [13]. To support this increase in customer service levels, support technology has become more sophisticated, involving elements of knowledge management and artificial intelligence. Nevertheless, the basic functions of the RUPERT system established in 1988 at Glaxo remain largely the same. Over and above this, however, as Peppard [14] has noted, the role of people skills and capabilities in delivering a successful Service Desk operation remains critical. Despite advances in technology support systems, a fully automated helpdesk function remains some years away for most organisations, and people skills will remain critical in providing an effective Service Desk operation.

References

[1] M. Wynn, "The business benefits of pc office systems and end-user computing at Glaxo Pharmaceuticals, 1984 – 1988", Journal of Information Technology, vol 4 (1), pp 17-29, March 1989

[2] N. Kumbakara, "Managed IT Services: the role of it standards", Information Management & Computer Security, vol 16 (16/4), pp.336-359, 2008.

[3] U. Sood, " 7 Essentials for a Top-Performing IT Help Desk", GCN Executive Roundtable Cloud, March 30, 2017. Available from: https://gcn.com/articles/2017/03/30/help- desk-essentials.aspx/

[4] D. Bridge and A. Dearden, "Knowledge Based Systems Support for Help Desk Operations: A Reference Model", Dept of Computer Scoiencwe University of York, April 1992. Available from http://citeseerx.ist.psu.edu/viewdoc/download?doi=10.1.1.47.5118&rep=rep1&type=pdf /2018.07.22

[5] L.Gonzalez, R. Giachetti, and G. Ramirez, "Knowledge management- centric help desk: specification and performance evaluation", Decision Support Systems, vol. 40, pp. 389-405, 2005

[6] A. Bryman and E. Bell, Business Research Methods, 3rd edition, Oxford: Oxford University Press, 2011.

[7] M. Saunders, P. Lewis, and A. Thornhill, Research methods for business students, 5th ed., 2009, England: Pearson Education Limited.

[8] M. Wynn and L. O'Callaghan, "Quality graphics at Glaxo Pharmaceuticals", Training Technology, vol. 1 (4), pp. 18-20, April, 1988

[9] R. K. Yin, Applications of Case Study Research. 3rd ed., 2012, London: SAGE Publications, Inc.

[10] R. Livesey, N. Moore, and D. James, User account policy. University of Gloucestershire, Library, Technology & Information Service, 2018.

[11] M. Wynn, "'Computerised training solutions at Glaxo", Interactive Learning International, vol. 4 (3/4), pp. 73-80, Sept 1987.

[12] The Stationary Office (TSO), ITIL Foundation Handbook, IT Service Management Forum, 2012.

[13] S.Serbest, Y.Goksen, O.Dogan, and A.Tokdemir, "Design and implementation of helpdesk system on the effective focuses of information system", Procedia Economics and Finance, vol 33, pp. 461.476, 2015

[14] J. Peppard "Managing IT as a portfolio of services" European Management Journal, vol. 21 (4), pp. 467–483, 2003.

Natural Language Processing in IBM Watson Assistant, an Automatic Verification Process

[1,2]Beatriz Consciência, [1,2]Simão Santos, [1]Pedro Martins, [2]Luís Pombo, [1]Steven Abrantes, [1]Cristina Wanzeller

[1]Polytechnic Institute of Viseu, Dep. of Computer Sciences, Viseu, Portugal
{pedromom, steven, cwanzeller}@estgv.ipv.pt
[2]Sofinsa, IBM, Portugal
{beatriz.consciencia, simao.santos, jose.luis.pombo}@pt.softinsa.com

***Abstract*—The exponential growth of Artificial Intelligence powered systems affect us all. IBM Watson Assistant is one of the central AI-powered systems used in small and larger business. For instance, in Portugal, IBM Watson is powering call-center systems for companies in banking and telecommunications business. This paper proposes a strategy that can automate the verification process of generalization capability in the creation of chat-bots using IBM's platform. K-Fold cross-validation is a favorite technique in machine learning for estimating the performance of a learned hypothesis on a data set. Therefore, the proposed method is not new for testing. However, this method is newly applied to the chat-bot application using IBM's platform. In this paper, the primary goal is to make the chat-bot testing process automated, with the objective of making it faster, more productive, and efficient. Algorithms like k-fold cross-validation demonstrate the need for a representative and reasonable amount of data when it comes to training IBM Watson in his ability to learn.**

Keywords-IBM Watson; cognitive computing; neural networks; natural language processing.

I. INTRODUCTION

As said, the progress in artificial intelligence is incredibly fast, close to exponential [13]. Companies are adopting this technology so they can make the most of it. For instance, call centers, hotel's room services, car's dashboards or even information balconies in airports are being redesigned to embrace AI.

In 2010, IBM presented Watson, a supercomputer whose architecture was question-answer based [15]. Recently, in March 2018, IBM launched IBM Watson Assistant, a platform that allows every user to create his very own chat-bot within the domain of knowledge that the user intends the chat-bot to be fluent at. Using this platform, chat-bots are built and deployed in a few clicks in a very user-friendly graphical interface. For that matter, users only need to acknowledge few terms and concepts such as "intent", "entities", "dialogue" and "digressions". In the upcoming section, those concepts are described.

With IBM Watson Assistant, users can deploy their chat-bots on other platforms such as Slack or even Facebook. However, when building a chat-bot application for an enterprise, it needs to be tested before going into production phase and, that testing should be carried out by real-life users who interact with it and point out the failures on recognition and misunderstandings of user intentions. The main goal is to make chat-bot testing an automated process so that it becomes a lot faster and consequently more productive and efficient for the business.

This research work presents a solution for the automation of the testing process in the creation of chat-bots in the Portuguese language, with the IBM Watson Assistant platform.

This paper is organized as follows. Section II, makes a brief resume of the related work. Section III, describes basic background knowledge regarding chat-bots. Section IV explains, making use of some examples how the k-fold algorithm is applied. Section V, identifies the case study used to motivate this work. Section VI, briefly describes, work-in-progress, implementation. Finally, section VII, point out some of the conclusions and future work directions.

II. RELATED WORK

IBM Watson Assistant was presented on March 19, 2018, at the "IBM Think 2018" in Las Vegas. It is a branch of the giant IBM Watson aimed at creating assistants intuitively [6].

Virtual assistants created with the help of Watson Assistant are targeted for business environments, which means that its main ambition is not to be available directly to the general public. [4]. Watson Assistant has already been tested in the new Maserati GranCabrio dashboard computer and at the Munich airport, integrated with a robot, destined to answer to transient issues [2].

Besides, the IBM Watson Assistant, as stated earlier, is possible to integrate into any and every business [3]. For example, in a hotel, this can be integrated into each room and allow the customer through simple commands to control the temperature and ambient lights or even ask for room services [5]. Another example, in the case of a bank, makes it possible to integrate IBM Watson Assistant as a chat-bot on the institution where the customer can carry out operations without the need of replacing the concept of "customer support" [10]. However, IBM Watson Assistant, as well as IBM Watson, do not communicate in Portuguese of Portugal. There is not yet an algorithm of natural language comprehension oriented to Portuguese, given the complexity and ambiguity of the language.

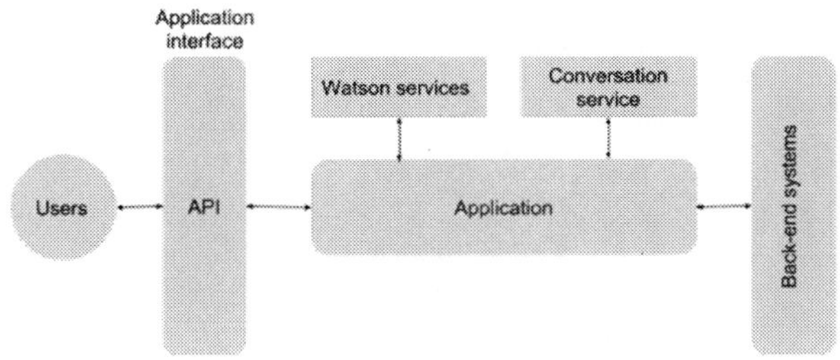

Fig. 1: IBM Watson Assistant integration

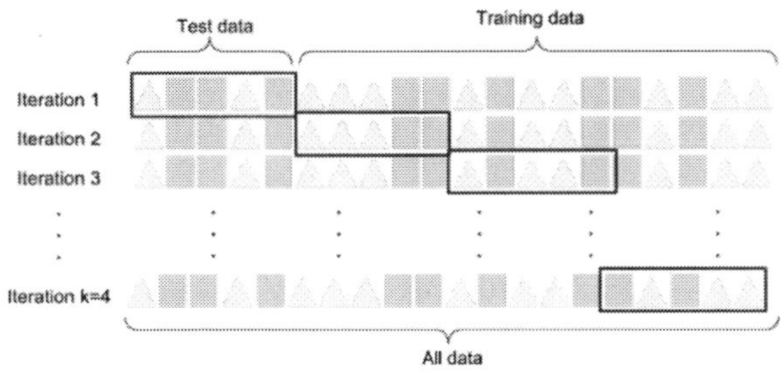

Fig. 2: Samples Partition into k equal sized sub-samples. Image adapted from [14].

Watson is also an assistant based on an intent architecture [16], where intents represent the user's intention in his or her speech, that is, what he or she desires and can usually be identified by a verb like "to go" or "to do" [12].

To the best of our knowledge, in the field of Portuguese (Portugal) chat-bots, this area is uncharted.

The Portuguese language, today is the sixth most spoken language in the world, with 244 million speakers spread across five continents. Thus, there is a large market for the Portuguese language that has not been yet conquered by AI-powered chat-bots.

Watson Assistant proposes that its users will be able to, in a short time, create a chat-bot, with which they can establish a dialogue. That creation is done in three steps, corpus creation, design, and integration [7]. Figure 1 shows a scheme of IBM Watson Assistant integration.

III. Concepts of chat-bots architecture

As said before, some concepts need to be understood before diving into chat-bot development. Most of those concepts are regarding "intents" and "entities." Intents represent the intention of a user when he/she addresses a chat-bot. For instance, if someone says "I want to buy a book", wanting to buy was the intention and therefore it would trigger an intent that contains other phrases related to that intention, and that can also trigger it. Another key term concerning to chat-bots is "entity". Referring again to the example sentence, "I want to buy a book", the entity, in this case, would be "books". So, entities are words that put intents into perspective giving them a purpose.

The other two terms that are relevant when it comes to developing a chat-bot with the IBM Watson Assistant platform are "dialogue" and "digressions". The first term refers to the dialogue flow which the chat-bot need to follow to have a structured speech. The second term, "digressions", refers to the conditions which allow the user off the chat-bot to jump from one intent to another and return to the former one if intended. Those represent a set of rules that allow the chat-bot to change context without losing the sense of the speech.

IV. K-fold Cross Validation

K-fold Cross Validation is algorithm mainly used in machine learning related problems where the main goal is to predict something and/or to estimate the reliability of a predictive model [1]. It is mostly applied to Supervised Learning models where a data-set is given with known labels of the target, on which training is performed (training data-set) [9]. It is also given a data-set of unlabeled data, against which the model is tested (testing set) [8]. The purpose of cross-validation is to test the model's ability to predict new data that were not used in estimating it, to flag problems like over-fitting and to give an insight on how the model will generalize when deployed on production [17].

To test and validate the proposed system, the original sample is randomly partitioned into k equal sized sub-samples, as can be seen in Figure 2. Of the k sub-samples, a single sub-sample is retained as the validation data for testing the model, and the remaining $k-1$ sub-samples are used as training data. The cross-validation process is then repeated k times (the folds), with each of the k sub-samples used exactly once as the validation data. The k number of results from the folds can then be averaged to produce a single estimation. The advantage of this method over repeated random sub-sampling is that all observations are used for both training and validation, and each representation is used for validation precisely once [11].

Given the example "I want to buy a book", that was previously classified as related to specific intent, the k-fold algorithm will match that previous classification with the classification resulting from IBM Watson Assistant. After that, it will present the result of that comparison (equal or not equal). This approach is then applied to the entire data model (i.e., the test array). When all instances have been compared, the k-fold algorithm will provide the results for each metric (e.g., Accuracy) per k fold which can be used to generate synthetic parameters such as the average accuracy of the whole chat-bot.

V. Case Study

For this study, a chat-bot was developed in IBM Watson Assistant platform. That chat-bot was a banking related one, therefore named Credit-bot.

Credit-bot is a chat-bot created just for this study on an academic proposal. It allows performing simple tasks that usually require a trip to the local bank branch. For instance, making a bank wire transfer; open an account, or even making a credit simulation. Notice, again, this tasks were simulated for academic purposes only.

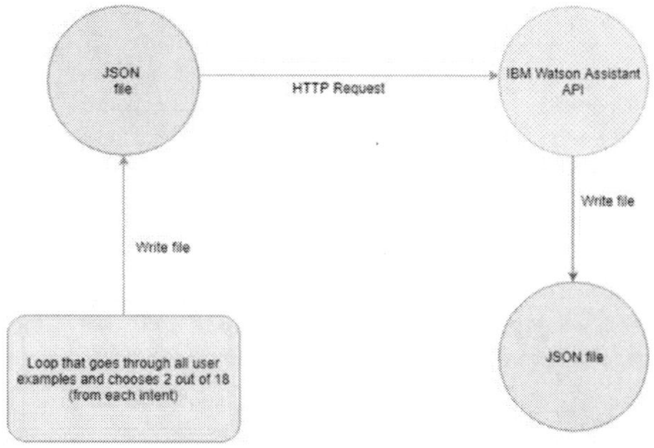

Fig. 3: Implementation process

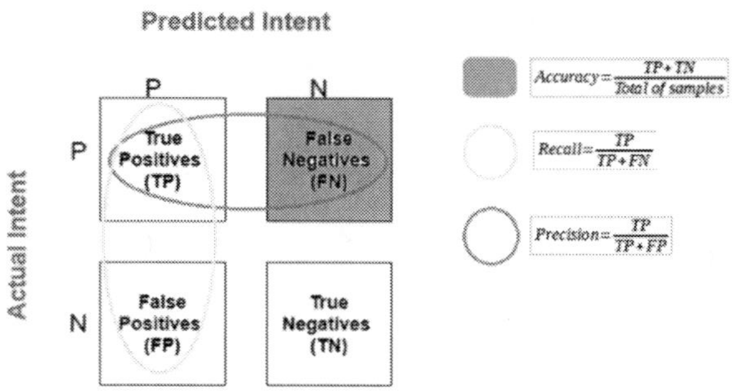

Fig. 4: Metrics

To make Credit-bot learn information about this domain some intents where created such as "Making Transfer" that compiles a set of useful user examples. User examples are phrases the user may address to the chat-bot and which need to relate to the respective intent to follow a workflow. Eleven intents to perform banking related actions were created, and twenty user examples were given to each intent, making a total of 220 user examples. Besides that, five entities were also created containing female names, male names, Portuguese city's, villages, terms related to account types and credit types.

This case-study it is aimed to create a chat-bot able of answering in Portuguese (Portugal) natural language to most frequent users requests. For instance, when using a chat, the user can authenticate himself, and then request in natural language (written) to transfer some money from his account to any other account. The same way, he can request credit, loans, pay regular bills, or even create routine tasks.

VI. Implementation

After the chat-bot creation in IBM Watson Assistant platform, it was produced a JSON file with the user examples needed to feed the chat-bot. This JSON file was then sent to the IBM Watson Assistant platform that reads the file and trains the chat-bot. For this study, the 20 examples per intent were hand-crafted (in a total of 220 examples). For each intent, the similar examples were split, where 18 were assigned to training data and 2 for testing. The process of sending this arrays to IBM platform is also performed through a JSON file. The process was then repeated ten times, as ten is the number of k. All samples were then sent to the IBM Watson Assistant through the API, where they were classified. This process scheme can be seen in Figure 3.

As a result of the classification, the API outputs a classification JSON file that was used by the k-fold algorithm. That JSON file was the target of an examination by the algorithm itself to compare the classification made by the IBM Watson Assistant versus the expected rating. That comparison resulted in another classification in another JSON file being the parameters "True Positive" when the classification made by the system was equal to the expected, and "False Positive" when it was not. Moreover, when an example was classified with "True Positive" the other examples must be classified as "True Negative", and when an example was classified as "False Positive" the other must be classified as "False Negative".

That classification was then used to calculate parameters like accuracy, recall, precision, and F1 score. This classification and obtained results are based on a confusion matrix. The method used to reach the metrics are depicted in Figure 4. The experimental results can be seen in Table I.

The IBM Watson Assistant platform was then powered with a set of test arrays. Each test array consisted on a set of examples, frequently named "user examples", that are phrases that a user might address to IBM Watson Assistant to trigger an intent. Each test array had in total 22 user examples. The test arrays were then used to test the platform generalization ability when classifying examples that were unknown. This test was performed to analyze where the model fails, so that model gaps are recognized and corrected. Gaps include, for example, spelling errors, unstructured sentences or even slang words, are translated into intents (e.g., if the user addresses to IBM Watson the sentence "send money" the system must recognize it with the intent "Make a transfer") that require a greater quantity or quality of data for disambiguation.

Results in Table I, show that it is possible for the proposed model to identify almost all true positives (TP). Thus, having a high Recall value (close to 1). However, there is also a high number of false positives (FP), which tells us that in many cases the platform classification fails, assuming the test cases as being relative to an intent of which they are not.

As for the means of precision by intent, these tell us that two intents have more miss-classification by the platform, such as attempting to "AffirmativeResponse" (positive answer) and "MakeTransfer" (make transfer). These intents are therefore more susceptible to ambiguity than the rest, which may lead to a dialogue with many doubts, making it unpleasant to the user. One possible reason for the ambiguity is the similarities in the training-set.

As far as the averages by k are concerned, from the same statistical distribution, presented values are equal, independently of the data subset, allowing no other conclusions than the one obtained.

TABLE I: RESULTS FROM METRICS

INTENT	AVERAGEPRECISION	STDPRECISION	AVERAGEACCURACY	STDACCURACY	AVERAGERECALL	STDRECALL	AVERAGEF1SCORE	STDF1SCORE
AFIRMATIVERESPONSE	0.4	0	0.945	0	1	0	0.571	0
MAKETRANSFER	0.4	0	0.945	0	1	0	0.571	0
FAREWELL	0.417	0.029	0.947	0.003	1	0	0.588	0.028
CAPACITIES	0.438	0.048	0.949	0.004	1	0	0.608	0.046
BOT_CONTROL_CHANGE_SUBJECT	0.45	0.05	0.95	0.005	1	0	0.619	0.048
BOT_CONTROL_CLARIFICATION	0.458	0.049	0.951	0.004	1	0	0.627	0.047
CREDITSIMULATION	0.464	0.048	0.951	0.004	1	0	0.633	0.045
ACCOUNTBALANCE	0.469	0.046	0.952	0.004	1	0	0.637	0.044
NEGATIVERESPONSE	0.472	0.044	0.952	0.004	1	0	0.64	0.042
OPENACCOUNT	0.475	0.042	0.952	0.004	1	0	0.643	0.04
GREETING	0.477	0.041	0.952	0.004	1	0	0.645	0.039

Main conclusions point out that "AffirmativeResponse" and "Make a transfer" has a lower generalization capability when compared to other intents. As stated above, this is because the given examples are very similar.

VII. Conclusions

The main goal of this paper was to outline a strategy to test the quality of a chat-bot automatically. This was achieved through means of the application of the k-fold cross-validation algorithm on the outputs of Watson Assistant intent classification, which allow us to assess the generalization ability of the underlying model through multiple metric which, in turn, give us a strong indication on the quality of the chat-bot solution. In practice, the advantages include the ability to identify in which topics the chat-bot will have more trouble figuring out the user true intent (therefore correcting the situation by adding more relevant data or splitting intents) and a way to continuously improve the quality of the chat-bot by continuously providing real-world usage data while measuring the quality improvements of the model. For instance, the chat-bot had less generalization ability towards the intent "MakeTransfer". This can be concluded through the values of average accuracy, that for that intent are "0", meaning that there was no record of true positive neither true negatives which means that there has the only record of false negatives and false positives. There is, none user example trigger the intent correctly. This is related to the quality of data given to the chat-bot since example phrases related to the entities do not variate much in content. Therefore we must find other ways to invoke the intent and consequently feed the intent with more examples so it can generalize better. This also happens with the intent "AffirmativeResponse".

Future work will include semantic analysis, integration of grammar parsing tools, text and/to speech recognition and other business tools provided by IBM Watson.

Other future work directions, lead to apply the proposed system to help visually impaired people, as well as, hearing impaired people.

Acknowledgements

"This article is financed by national funds through FCT - Fundação para a Ciência e Tecnologia, I.P., under the project UID/Multi/04016/2016. Furthermore, we would like to thank the Instituto Politécnico de Viseu for their support."

More, we especially would like to thank the Viseu Polytechnic professors: Prof. Cristina Wanzeller for her outstanding work supervision; Prof. Pedro Martins, for his support and orientation. Finally, we thank Softinsa, IBM group, for making possible this research work.

References

[1] M. W. Browne and R. Cudeck. Single sample cross-validation indices for covariance structures. *Multivariate behavioral research*, 24(4):445–455, 1989.

[2] H. Chen, R. H. Chiang, and V. C. Storey. Business intelligence and analytics: from big data to big impact. *MIS quarterly*, pages 1165–1188, 2012.

[3] Y. Chen, J. E. Argentinis, and G. Weber. Ibm watson: how cognitive computing can be applied to big data challenges in life sciences research. *Clinical therapeutics*, 38(4):688–701, 2016.

[4] L. Comerford, D. Frank, P. Gopalakrishnan, R. Gopinath, and J. Sedivy. The ibm personal speech assistant. In *Acoustics, Speech, and Signal Processing, 2001. Proceedings.(ICASSP'01). 2001 IEEE International Conference on*, volume 1, pages 1–4. IEEE, 2001.

[5] F. Curbera, M. Duftler, R. Khalaf, W. Nagy, N. Mukhi, and S. Weerawarana. Unraveling the web services web: an introduction to soap, wsdl, and uddi. *IEEE Internet computing*, 6(2):86–93, 2002.

[6] R. High. The era of cognitive systems: An inside look at ibm watson and how it works. *IBM Corporation, Redbooks*, 2012.

[7] P. Imrie and P. Bednar. Virtual personal assistant. In *ItAIS 2013*. AIS Electronic Library (AISeL), 2013.

[8] A. Kapur, A. Kapur, N. Virji-Babul, G. Tzanetakis, and P. F. Driessen. Gesture-based affective computing on motion capture data. In *International conference on affective computing and intelligent interaction*, pages 1–7. Springer, 2005.

[9] R. Kohavi. Glossary of terms. *Special issue on applications of machine learning and the knowledge discovery process*, 30(271):127–132, 1998.

[10] A. P. Massey, M. M. Montoya-Weiss, and K. Holcom. Re-engineering the customer relationship: leveraging knowledge assets at ibm. *Decision Support Systems*, 32(2):155–170, 2001.

[11] G. McLachlan, K.-A. Do, and C. Ambroise. *Analyzing microarray gene expression data*, volume 422. John Wiley & Sons, 2005.

[12] S. Memeti and S. Pllana. Papa: A parallel programming assistant powered by ibm watson cognitive computing technology. *Journal of Computational Science*, 2018.

[13] M. Minsky. Steps toward artificial intelligence. *Proceedings of the IRE*, 49(1):8–30, 1961.

[14] F. Pedregosa-Izquierdo. *Feature extraction and supervised learning on fMRI: from practice to theory*. PhD thesis, Université Pierre et Marie Curie-Paris VI, 2015.

[15] C. Thompson. What is ibm's watson. *New York Times Magazine (accessed on October 2018). http://www. nytimes. com/2010/06/20/magazine/20Computer-t. html*, 2018.

[16] C. Todd, R. Vazquez Pena, and R. Srinivas. Evaluation of artificial intelligence frameworks. *SMU Data Science Review*, 1(1):10, 2018.

[17] X. Xie, J. W. Ho, C. Murphy, G. Kaiser, B. Xu, and T. Y. Chen. Testing and validating machine learning classifiers by metamorphic testing. *Journal of Systems and Software*, 84(4):544–558, 2011.

Evaluating a User Story Based Recommendation System for Supporting Development Processes in Large Enterprises

Maria Lusky, Matthias Jurisch, Stephan Böhm

Faculty of Design – Computer Science – Media
RheinMain University of Applied Sciences
Wiesbaden, Germany
Email: {maria.lusky, matthias.jurisch, stephan.boehm}@hs-rm.de

Katharina Kahlcke

User Experience Consulting
DB Systel GmbH
Frankfurt, Germany
Email: katharina.kahlcke@deutschebahn.com

***Abstract*—In mobile application development projects, large enterprises have to face special challenges. To meet these challenges and to meet today's high expectations on user centered design, inter-project knowledge transfer of software artifacts becomes an important aspect for large software development companies. For supporting this kind of knowledge transfer, we propose an approach for a recommendation system based on textual similarity of user stories that are computed via standard information retrieval techniques. We also present a three-step evaluation of this approach, comprising data analysis, a survey and a user study. The results tend to support our approach and not only show that user story similarity rated by users and rated by an algorithm correlates, but also demonstrate a strong relation between user story similarity and their usefulness for inter-project knowledge transfer.**

***Keywords*–Mobile Enterprise Applications; User Stories; Recommendation Systems; User Centered Design.**

I. INTRODUCTION

In recent years, the user centered design approach has become an integral part of software development, and also for mobile application (app) development. Often, at the beginning of an agile development process, requirements for an app are analyzed and are written down in the form of user stories. They are short requirement descriptions from the user's point of view. Based on these user stories, during the further development process other software artifacts, such as documentation, screen designs, or source code are created to support the development process. Especially in large enterprises, the reuse of these software artifacts can save time and resources, since large software development companies are facing several challenges: There are multiple development projects at the same time, resulting in a large number of software artifacts. Due to a lack of time, these artifacts are often not properly documented in order to support a reuse of these materials. Furthermore, team members often do not know if there is a project with similar requirements and which coworker they can contact about a reuse of software artifacts. In general, large software development companies deal with lack of transparency in development projects, contact persons, and software artifacts.

Saving time and resources through reuse is especially desirable for enterprises in the Mobile Enterprise Application (MEA) market: due to digitalization trends, these enterprises are developing many apps for various customers at the same time [1]. Nevertheless, quick time to market is important because of a rapidly changing mobile ecosystem. Additionally, enterprises can only access a limited number of specialists that should focus on demanding tasks and work on difficult problems that have not been solved in the company already. This is the context of the project PROFRAME [2], that tries to support the reuse in software project using several approaches. Given this background, in this paper we propose an approach that supports the reuse of software artifacts in mobile app development projects based on textual similarity of user stories. In a previous paper on this problem, we showed that similar user stories can be identified via classical methods of information retrieval [3]. In the following evaluation, we investigate how well these methods work in a real world dataset and how useful our approach is for employees of a large software development company. Therefore, Section II provides an overview on related work. Section III introduces our approach. An evaluation is described in Section IV and its results are presented in Section V. These results are discussed afterwards in Section VI. Section VII concludes this paper and gives an outlook on further research.

II. RELATED WORK

Supporting reuse in the context of software development can be facilitated in many ways. One of these ways is best practice sharing, where good solutions for common problems are shared. However, this requires a lot of work and time to find common problems and respective solutions. Especially in the context of MEA development, time is an important factor. Therefore, supporting this process with automated approaches seems to be promising. An automated approach in this area is the use of recommendation systems [4]. Recommendation systems recommend items to users based on item similarities or preferences of similar users. This idea can be applied to software engineering, where a recommendation system can recommend software engineering artifacts to developers [5]. The goal of these recommendation systems is mainly to support the software development process, especially focused on the implementation phase and fixing bugs.

An important field in this area is issue triage. This field deals with supporting the management of bug reporting systems. This includes both recommending specific developers for a given bug report, as well as detecting duplicate bugs. Usually, standard information retrieval techniques or shallow machine learning approaches are used [6]: An approach by Runeson et al. [7] is based on information retrieval techniques and tries to detect duplicates. Other approaches build on including other information besides text, e.g., execution information [8]. In [9], a framework for building recommendation systems for issue triage is presented. This is done by linking both developers and bug reports to project components. Besides using classical information retrieval methods, more recent approaches use deep learning-techniques like word embeddings [10]. While this area also includes computing similarities between textual descriptions in the area of software development, there are some important differences: (1) bug reports are often written from a very developer-centric perspective. (2) They usually contain a lot of technical information like log output. (3) The main goal of issue triage is not to support reuse, but to support bug management tasks.

Another approach to improve the knowledge management in software development projects is to document and store project information in an accessible way, e.g., in architectural knowledge management tools [11]. These approaches have also been applied in industrial case studies and were deemed fit for usage in an industry context: [12] evaluated a semantic architectural knowledge management tool that is based on existing data on software design patterns and their application in software projects. However, if this kind of data is not already available the overhead for documenting usage of design patterns can be too high for an application in practice. This is especially an issue for the fast-paced market of mobile enterprise applications.

In the last decade, user stories as a user-centric representation of requirements were introduced [13]. A typical user story is at most two sentences long and consists of a reference to a certain type of user, a description of an activity this user wants to do with the software and a reason why this will help the user. As an attachment to the user story, acceptance criteria add more detailed information to the user story. Only few approaches exist to support software reuse in the context of user stories: [14] proposes a recommendation system based on user stories and evaluates this system on a project history. However, it is not clear how helpful these recommendations would be when actually working on a new project. In our previous work [3], we evaluated how well information-retrieval-based approaches can distinguish between two types of user stories and which aspects of the user story are important to it.

In this paper, we expand this previous work to an evaluation of how useful recommendations on a real-world software engineering dataset are and what information needs to be contained in these recommendations to make them actually helpful. This issue has not been addressed by the approaches we mentioned in this section and is required to make recommendations in the context of user stories usable in practice. The only way to work on this issue is to conduct a survey with practitioners from the industry.

III. Recommendation Approach

Our general approach to supporting reuse is to use established techniques from information retrieval in order to recommend textually similar user stories to the story a participant in an app development-project is currently working on. The information that is attached to the recommended user stories (e.g., screen designs, textual documents or source code) can then be used to support current efforts. In this way, team members could reuse results from different projects without previously knowing about these projects.

To find textually similar user stories, a search based on the well-known information retrieval approach Term Frequency-Inverse Document Frequency (TF-IDF) and stop word removal is used. Stop words are words that occur frequently in texts so that they do not contain useful information. Examples for stop words are "I", "the", "a", etc. These words are removed from the user stories before processing the user stories with TF-IDF, which represents texts as follows: Each document d (i.e., a user story) is represented by a vector w_d, that contains an entry for each term used in the dataset. Each vector component $w_{d,t}$ represents the importance of a term t for the document d. This representation is computed by the frequency of a given term in the document $tf_{d,t}$ multiplied by the inverse document frequency $\log \frac{N}{df_t}$, where N is the number of all documents and df_t is the number of occurrences for a given term in all documents. This yields the following formula for a document's vector representation:

$$w_{d,t} = tf_{d,t} * \log \frac{N}{df_t}$$

To compute the similarity between documents, the cosine of the angle of two vectors is used. The naive approach for similarities would be to compute the euclidian distance between vetors, however this would favour documents with similar lengths.

Cosine similarities are then used to order texts regarding their relative similarities. Thus, we do not use similarity scores as an absolute value, but only to distinguish between more and less similar documents. To find similar user stories to one given user story, the similarity is computed according to the described procedure. User stories are then ordered by their similarity and the user story with the highest score is considered the most similar.

IV. Evaluation

We evaluated the recommendation approach described in Section III and the importance of using recommendation systems for reuse in development processes in large enterprises by investigating the following research questions:

1) Which kind of knowledge transfer is already being practiced?
2) Can an automated recommendation system be useful for supporting knowledge transfer?
3) Is there a relation between user story similarity and their usefulness?

A. Methodology

To answer these research questions, we conducted an evaluation comprising three steps: In the first step, we analyzed a dataset of user stories from a large German software development company. In the second step, we distributed a questionnaire covering questions about practices in inter-project knowledge transfer in general. In the third step, we invited participants to single sessions where they were asked to solve tasks focusing on user stories in inter-project knowledge transfer.

The number of participants in this study was limited to a small number due to the testing requirements: all participants had to come from the same company with a specific expertise on the implementation of the user stories and as references for the similarity comparison. Thus, the results are far from representative and cannot be considered as an empirical validation. However, the results of this prestudy can five some critical and usable expert feedback on the potential usefulness and applicability of our approach.

B. Dataset

To evaluate the usefulness of recommendations based on user stories in the area of Mobile Enterprise Application Development, we used a dataset of real-world user stories out of nine app development projects provided by an industry partner. The dataset contains 591 user stories, of which 355 are long enough to contain meaningful information. User stories were considered long enough when they were at least 80 characters long, which is roughly two times the length of only the formal aspects of a user story description. This boundary was set by investigating example user stories. A histogram of story length (in characters) is shown in Figure 1. From the distribution of story length and the standard deviation, we can already conclude that the dataset is very heterogeneous, as could be expected in a real-world dataset. The data is not only heterogeneous regarding the textual length, but also regarding their specifity and their degree of abstraction. For example, some user stories describe fixing typos in data protection regulation informations, while others describe a high level view of a location-based service.

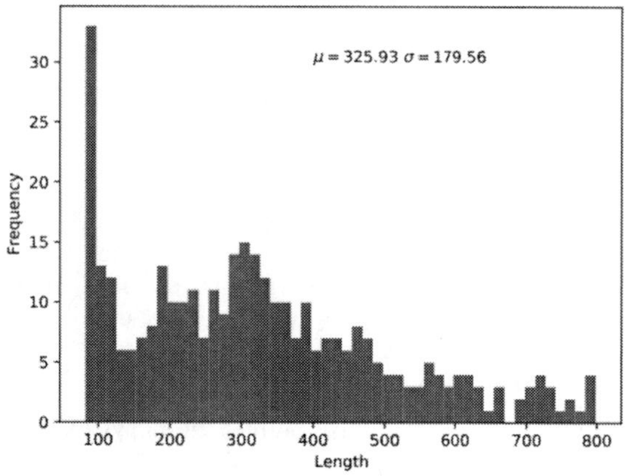

Figure 1. Distribution of User Story Length

For each user story in the dataset, we computed the top five most similar user stories according to TF-IDF, with cosine similarity both for stories from different projects as well as stories from the same project. A histogram of cosine similarity values between all user stories is shown in Figure 2. Stories in the same project are given higeher similarity values than stories from different projects, which indicates that it is possible to differentiate between projects using cosine similarities.

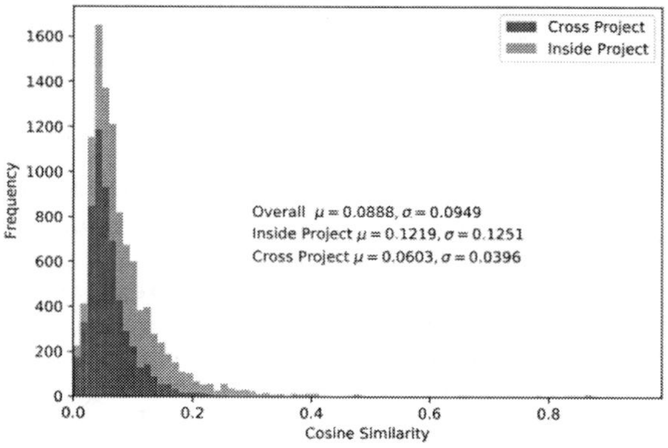

Figure 2. Distribution of User Story Similarities

C. Survey

To get an overall overview of the requirements in user story-centered reuse, we designed a questionnaire that comprised ten questions about inter-project knowledge transfer. The questionnaire was online for 17 days and was distributed among the employees of a large German software development company. It was also used for recruitment of participants for the user study. First, the participants had to specify their field of activity. We then explained to them our approach for inter-project knowledge transfer that underlay the questions, that is the re-use of software artifacts and knowledge, such as user stories, screen designs, documentation or source code, during development projects of mobile applications.

We asked the participants, how useful such a knowledge transfer is and in which way and how regularly it is already being practiced in their department. Further, they had to name obstacles that occur with inter-project knowledge transfer. Then, we asked them to rate the usefulness of particular software artifacts in this context and they had to assess the viability of such a knowledge transfer in their department. They were asked to rate the usefulness of a software that would support knowledge transfer. Lastly, the participants were asked to rate the importance of additional information to specific software artifacts on a five-point scale. Importance may differ from usefulness in certain situations, since it is used to prioritize between different potentially useful artifacts.

D. User Study

The user study was carried out in one-on-one sessions with employees of a large German software development company. Each session lasted 20 to 30 minutes. We selected three user stories for which related user stories were known to be in the dataset. For each of these *reference* user stories, we computed the most similar user stories from the different projects with varying levels of similarity: one user story that the algorithm listed most similar, one that it listed as medium similar, and one that it listed as less similar, which lead us to three user story groups, one for each reference story.

Based on the reference user stories, the participants were asked to solve three tasks. First, they had to rank the user stories obtained by the algorithm regarding their similarity to

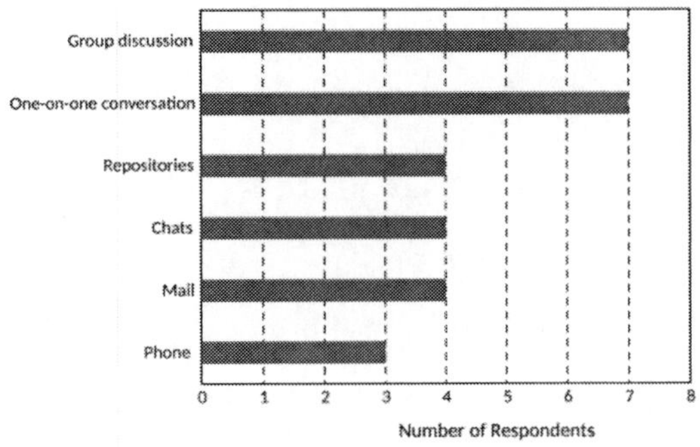

Figure 3. Currently Used Types of Knowledge Transfer

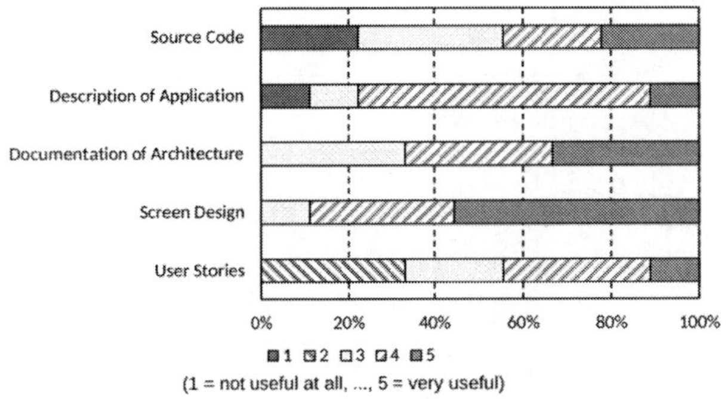

Figure 5. Perceived Usefulness of Artifacts

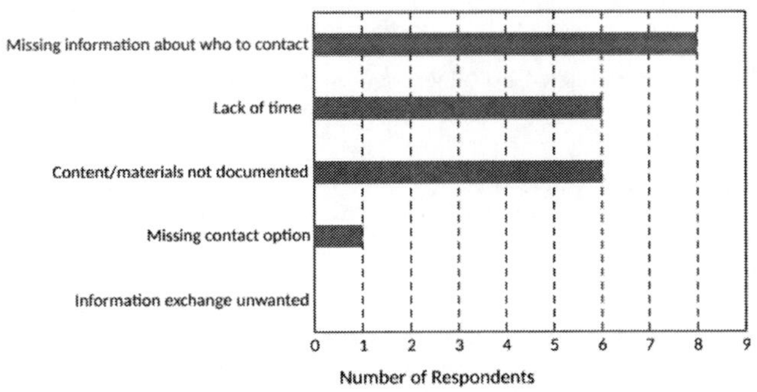

Figure 4. Existing Barriers for Knowledge Transfer

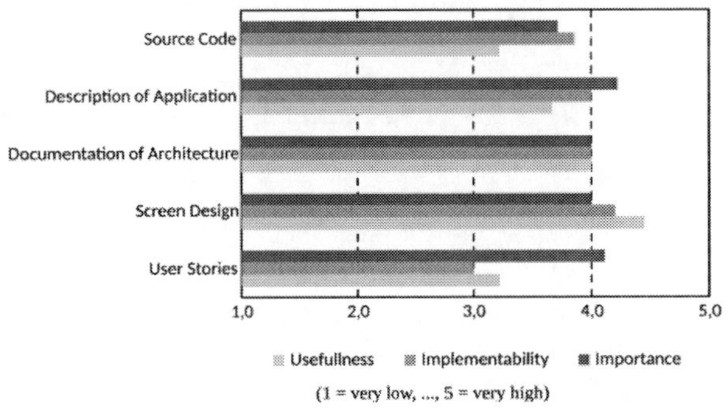

Figure 6. Responses on Potential Recommendation System Artifacts

the reference user story from the most similar to the least similar one. Then, they should rate the usefulness of each of the similar user stories. To determine the usefulness of the user stories, participants were told to estimate, how much artifacts (e.g., source code, design documents or documentation) produced during an implementation of a ranked user story could contribute to the implementation of the corresponding reference user story. Note that this is not included in similarity aspects: user stories can cover a roughly similar topic, however different levels of abstraction, different user types or platforms or technical aspects could make it impossible to actually reuse the results of the implementation of a user story in a different context. Such user stories would be considered similar by users, but finding these stories would not actually support reuse of artifacts related to one user story to another. Concluding the session, they were asked to name additional information that should be provided by the recommendation system in order to support the implementation of the reference user story.

V. RESULTS

The questionnaire was answered by nine employees of a large German Software development company. While nine participants are not enough to allow a detailed statistical analysis, this number is in general considered enough for usability testing [15]. Eight participants specified their field of expertise as conception, one as implementation.

All of the participants rated the knowledge transfer described by us (that is, the re-use of software artifacts and knowledge, such as user stories, screen designs, documentation or source code, during development projects of mobile applications), using a five-point scale, as useful or rather useful (median=5; maximum=5; minimum=4). The currently used types of knowledge transfer selected from a list of pre-made options are shown in Figure 3. All of the participants stated that they already practiced this kind of knowledge transfer, seven via one-on-one conversations or group conversations, four via e-mail, chats or by using knowledge bases, and three via phone calls. On average, each person practices three methods of knowledge transfer. Only one stated to practice it on a regular basis, and eight practice it as needed. Obstacles for knowledge transfer selected by participants from a list of possible obstacles are shown in Figure 4. The most often named obstacle was missing information about a contact person (eight), followed by missing documentation of content and materials and lack of time (six respondents each). The participants described further obstacles as being unaware of the existence of reusable materials, as well as not knowing where to look for information regarding reusable artifacts.

User ratings for usefulness of artifacts for Knowledge transfer on a five-point scale are shown in Figure 5. Screen designs were rated as most useful (median=5, maximum=5, minimum=3), followed by documentation of the software architecture (median=4, maximum=5, minimum=3). Ratings for potential usefulness, implementability and importance are shown in Figure 6. Regarding the viability of such a knowledge transfer in their department and in relation to specific

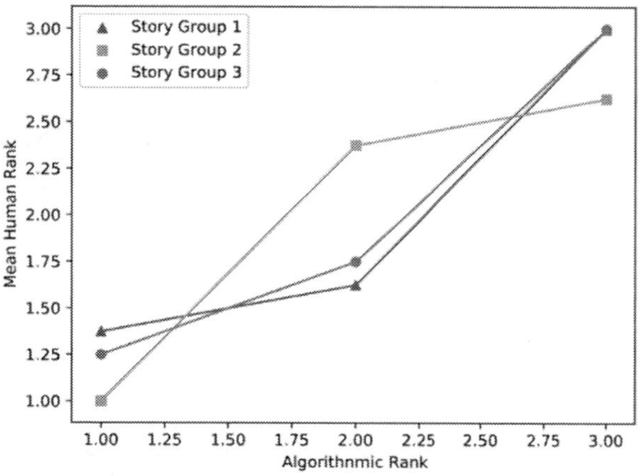

(a) Ranking of the user stories by the algorithm and mean ranking by the participants.

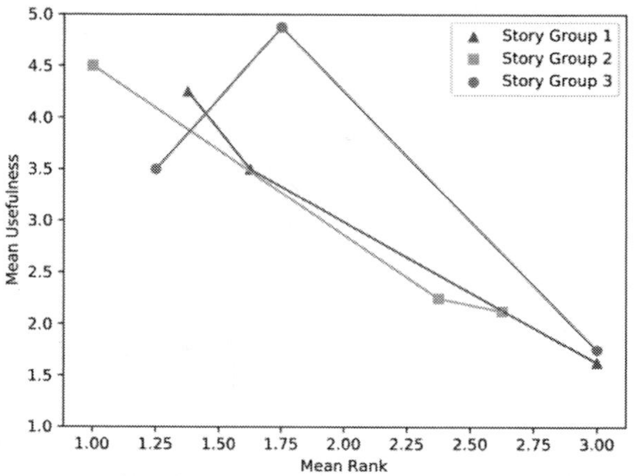

(b) Mean estimated usefulness and mean ranking by participants per user story.

Figure 7. User study results

software artifacts, the highest implementability was considered for screen designs, followed by documentation of the software architecture and use case descriptions. Furthermore, the answers revealed that most knowledge transfer that is already practiced concerns screen designs (done by 4 participants), documentation of the software architecture (2 participants) and user stories (1 participant).

The participants rated a software solution for supporting knowledge transfer on a five-point scale as rather useful (median=4; maximum=5; minimum=1), with 6 participants considering it rather useful or useful. Regarding the importance of additional information, information on use case descriptions were rated as most useful (median=4; maximum=5; minimum=3). In general, any kind of additional information (e.g., source code, screen designs, architecture documentation) was rated as "rather important" for all kinds of software artifacts.

Of the eight participants of the user study, all were working in conception respectively design. Results of the user study are shown in Figure 7. Results of the first task show that the participants ranked the user stories similar to the ranking of the algorithm. Figure 7a shows the ranking of the result user story similarity to the reference story by the algorithm and the mean ranking by the participants. The data shows that user story similarity of the algorithm seems to resemble the perceived user story similarity by humans: The three user stories ranked as most similar by the algorithm also got the highest similarity rankings by the participants. The user stories ranked as second by the algorithm were partially ranked as more and partially as less similar, but in general reflect the algorithmic ranking. For stories that are not obviously the least or most similar, this result was to be expected. Accordingly, the three least similar user stories of the participants match those ranked by the algorithm.

Further on, for each user story in the three user story groups, we calculated the mean similarity ranking to the reference story given by the participants and the mean usefulness rating, according to the rated usefulness of a computed user story for the implementation of the reference story. As Figure 7b shows, there are two groups of user stories that are delineated from each other. The first group has higher usefulness values (3,50 to 4,88) and higher similarity rankings (1,00 to 1,75), while the second group has lower usefulness values (1.63 to 2.25) and lower similarity rankings (2,38 to 3,00). However, the user story with the highest usefulness (4,88) is ranked with medium similarity (1,75), while the two user stories with the lowest usefulness (1,63 and 1,75) are ranked as least similar.

VI. Discussion

The results of the questionnaire show that, in general, people appreciate knowledge transfer and that our approach for it meets the users' needs. This is also reflected in the fact that 7 of our participants already practice this kind of knowledge transfer. Although, direct contact and personal conversations are the preferred ways of doing so. Electronic ways for contacting each other are rarely used. One reason for that might be that electronic methods, such as emails, chats or knowledge bases, do not meet the user needs for knowledge transfer. Nevertheless, for all these methods the user needs to know, which person can be contacted for further information – our approach takes this important feature into account, that was also the most often named obstacle for knowledge transfer. The second obstacle is insufficient documentation – this problem is also taken account of in our approach, since it simplifies documentation by searching existing software artifacts based on similarities. In general, the answers confirm that our approach addresses the right issues and helps to eliminate the obstacles for knowledge transfer that have been named by the participants. The software tool proposed by us was said to be most useful for screen design. This answer is not unusual, since almost all participants of our questionnaire work in design and conception. However, screen design is their main field of activity and thus, we consider it positive that our approach is rated as useful for this field. Further, the viability was rated highest for screen designs. These results give some evidence that our approach can create additional value in one of the most important fields for knowledge transfer. All in all, two thirds of the participants consider our approach useful or very useful.

The results of the user study show that the user story similarity of the algorithm seems to be connected to the perceived human user story similarity. Furthermore, user story similarity mostly coincides with their usefulness. The data indicates that a low similarity implies a low usefulness. However, the most similar stories are not always the most useful ones, but to some extent a high similarity seems to be connected to higher usefulness. This confirms our assumption that similarity between user stories and usefulness for the implementation of another user story are not necessarily the same.

VII. Conclusions and Further Research

In conclusion, the evaluation provided valuable findings, so that our research questions can be answered as follows:

1) Which kind of knowledge transfer is already being practiced?
   Mainly, knowledge transfer takes place in personal conversations between two people and groups. It is not carried out on a regular basis, but as required. Our results suggest that everyone does practice knowledge transfer in one way or the other.
2) Can an automated recommendation system be useful for supporting knowledge transfer?
   Our results show that an automated recommendation system is a useful tool for supporting knowledge transfer, especially for screen designs.
3) Is there a relation between user story similarity and their usefulness?
   The results of our user study indicate a relation between user story similarity and their usefulness. Further, there also is a connection between user story similarity rated by an algorithm on the one hand and humans on the other hand.

As a next step, another iteration of the evaluation could be made in different companies, in order to receive results that are applicable in several contexts of work. Also, more approaches for computing similar user stories could be evaluated: A comparative study of textual similarity approaches such as word movers distance or taking metadata into account, would provide valuable insights.

Acknowledgment

This work was funded by the German Federal Ministry of Education and Research, grant no. 03FH032PX5; the PROFRAME project at RheinMain University of Applied Sciences. All responsibility for the content of this paper lies with the authors.

References

[1] A. Giessmann, K. Stanoevska-Slabeva, and B. de Visser, "Mobile enterprise applications–current state and future directions," in System Science (HICSS), 2012 45th Hawaii International Conference on, Jan 2012, pp. 1363–1372.

[2] M. Jurisch, B. Igler, and S. Böhm, "PROFRAME: A Prototyping Framework for Mobile Enterprise Applications," in CENTRIC 2016, The Ninth International Conference on Advances in Human-oriented and Personalized Mechanisms, Technologies, and Services, 2016, pp. 7–10.

[3] M. Jurisch, M. Lusky, B. Igler, and S. Boehm, "Evaluating a recommendation system for user stories in mobile enterprise application development," International Journal On Advances in Intelligent Systems, vol. 10, no. 1 and 2, 2017, pp. 40–47.

[4] M. Robillard, R. Walker, and T. Zimmermann, "Recommendation systems for software engineering," IEEE Software, vol. 27, no. 4, 2010, pp. 80–86.

[5] D. Cubranic, G. C. Murphy, J. Singer, and K. S. Booth, "Hipikat: A project memory for software development," IEEE Trans. Softw. Eng., vol. 31, no. 6, Jun. 2005, pp. 446–465. [Online]. Available: https://doi.org/10.1109/TSE.2005.71

[6] A. Goyal and N. Sardana, "Machine learning or information retrieval techniques for bug triaging: Which is better?" e-Informatica Software Engineering Journal, vol. 11, no. 1, 2017, pp. 117–141.

[7] P. Runeson, M. Alexandersson, and O. Nyholm, "Detection of duplicate defect reports using natural language processing," Proceedings - International Conference on Software Engineering, 2007, pp. 499–508.

[8] X. Wang, L. Zhang, T. Xie, J. Anvik, and J. Sun, "An Approach to Detecting Duplicate Bug Reports using Natural Language and Execution Information," Proceedings of the 30th international conference on Software engineering, 2008, pp. 461–470.

[9] J. Anvik and G. C. Murphy, "Reducing the Effort of Bug Report Triage: Recommenders for Development-Oriented Decisions," ACM Transactions on Software Engineering and Methodology (TOSEM), vol. 20, no. 3, 2011, pp. 10:1–10:35.

[10] S. Mani, A. Sankaran, and R. Aralikatte, "Deeptriage: Exploring the effectiveness of deep learning for bug triaging," arXiv preprint arXiv:1801.01275, 2018.

[11] R. Capilla, A. Jansen, A. Tang, P. Avgeriou, and M. A. Babar, "10 years of software architecture knowledge management: Practice and future," Journal of Systems and Software, vol. 116, 2016, pp. 191–205.

[12] M. Sabou et al., "Exploring enterprise knowledge graphs: A use case in software engineering," in European Semantic Web Conference. Springer, 2018, pp. 560–575.

[13] M. Cohn, User Stories Applied: For Agile Software Development. Redwood City, CA, USA: Addison Wesley Longman Publishing Co., Inc., 2004.

[14] H. Pirzadeh, A. D. S. Oliveira, and S. Shanian, "ReUse : A Recommendation System for Implementing User Stories," in International Conference on Software Engineering Advances, 2016, pp. 149–153.

[15] J. Nielsen, "How many test users in a usability study?" https://www.nngroup.com/articles/how-many-test-users/, website. [retrieved: August, 2018], 2012.

A User Rights Concept for Semantic Media in Ambient Learning Spaces

David Bouck-Standen, Christina Eggert, Alexander Ohlei, Michael Herczeg
Institute for Multimedia and Interactive Systems (IMIS)
University of Luebeck
Luebeck, Germany
email: [bouck-standen, eggert, ohlei, herczeg]@imis.uni-luebeck.de

Abstract—**In our research project Ambient Learning Spaces (ALS), users interact with Semantic Media. Semantic Media consist of both plain media, such as still images, audio, video, text, or 3D objects, enriched by semantic relations and annotations describing content and inheritance structure of the media. These annotations allow classifying and arranging the media into semantic models. Semantic Media represent the information users create, collect and access, e.g., in the museum or school context of our research project. As media are accessed and used in various applications from the family of ALS applications, a user rights concept is required to manage access to these media. Usually such user rights have to be set explicitly.**

Keywords-Semantic Media; User Rights Concept; Implicit User Rights Setting;

I. INTRODUCTION

A single instance of plain media, such as a still image, audio, video, text or 3D object, represents information. Only when such media are integrated into meaningful information structures, these media become valuable in specific contexts [1]. We decided to build up such information structures as semantic models through semantic annotations. Thus, plain media enriched by semantic annotations will be referred to as *Semantic Media*. In our research project *Ambient Learning Spaces (ALS)*, body- and space-related human-computer interaction, as well as the concept of cross-device interaction (XDI) defines the conceptual foundation. For this purpose, we have developed a family of interconnected learning applications. Providing these applications we focus on self-directed learning with interactive media [2]. In our ALS environment, learners create and interact with such Semantic Media in school and museum contexts, using their own mobile devices [3]. In these contexts, users will typically be school students, teachers, museum visitors, staff, or curators. They interact with Semantic Media from mobile or stationary devices running applications from the ALS family, all interconnected through a framework of web-based services and a data repository, the *Network Environment for Multimedia Objects (NEMO)*. NEMO stores and handles all media within the ALS research project. The framework is running in multiple instances cloud-based, as well as on-site with our project partners, two schools and two museums. During their use of ALS applications, users' access to media is not limited to the media they create themselves. The philosophy of ALS is to share knowledge and engage in communicative and collaborative learning processes through these systems, where knowledge is expressed through Semantic Media [1]. This requires a user rights concept in order to manage access to these Semantic Media.

Apart from setting user rights explicitly, it appears that in a multi-device and multi-user ecology, like ALS is used in schools and museums, implicit settings of user rights are even more significant. In an initial approach, we develop a concept, as outlined below, in which, we use semantic annotations to equip each instance of Semantic Media with the required settings.

In Section 2 we regard related work, distinguishing our approach from those applying explicit user right. In Section 3, we outline a concept that focuses on the implicit setting of user rights, which reflects the users' contextualized use of the media in the context of its use. In Section 4, we conclude with a summary and outlook, as well as discussion stimuli.

II. RELATED WORK

Explicit user rights settings are required in various contexts and have been researched based on, e.g., inheritance, role hierarchies or in privacy-sensitive contexts, by Lu et al. [4]. User rights on media can be perceived similar to access permissions. Ferrara et al. [5] outline a concept to reduce annotation overhead by automatically inferring access permissions, which is also applicable to media, whereas Treu et al. [6] introduce a concept of granting implicit access to personal location information on basis of certain rules applied to the relations between users.

Distinguishing the concept presented in this contribution from the related work, we connect the user's explicit group-based user rights settings for each medium with implicit user rights. These user rights are derived from the users' utilization of the media in context of the applications, in which they create or interact with the media.

III. CONCEPT FOR IMPLICIT USER RIGHTS

To illustrate the concept for implicit user rights presented in this contribution for discussion, in the following we will use a school scenario. This is based on our concept of Semantic Media, which is annotated in NEMO (see Figure 1).

In ALS, users create new media with the *Mobile Learning Exploration System (MoLES)* [1], an ALS application for mobiles. In a typical setting, MoLES is used collaboratively by a group of 3-4 school students. On a field trip, multiple groups create and annotate media using MoLES in order to answer questions. These media are transferred to and stored inside NEMO. After the field trip

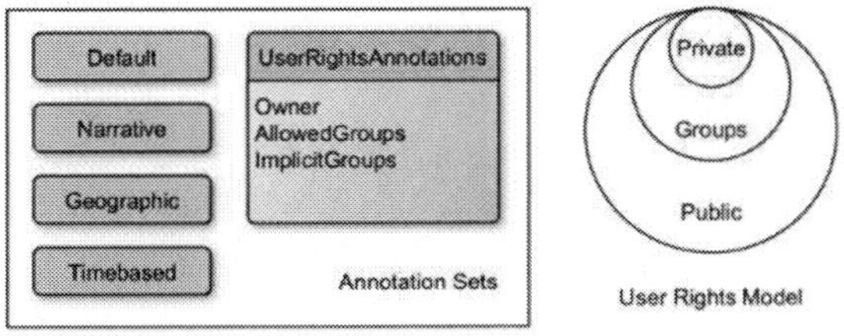

Figure 1. Left: NEMO annotation sets for Semantic Media. Right: Groups of the user rights model for ALS.

and back in school, the students prepare a presentation of their findings using the media they created previously. They discuss their findings and share media among each other and across their groups defined previously. When they finished their work, they showcase their presentation, at first in class. They can afterwards decide, whether to show their presentation (media) on displays in a public space in school.

The scenario illustrates the use of media in tasks like creation, collaborative usage, and sharing. As the scenario outlines, no user explicitly sets user rights to their media. Instead, they simply interact with the media through ALS applications. Through this interaction, they implicitly set the user rights on their Semantic Media.

This concept relies on a group-based model with three types of access permissions: *noaccess*, *read*, and *write*. These are sufficient for all ALS scenarios. At first, the users are explicitly assigned into *groups*, e.g., depending on classes, or school projects. They are also implicitly assigned to a group, e.g., when a group of students use MoLES together. Whenever a user in ALS creates media, NEMO applies a default *private* setting to protect the media from access of others (see Figure 1) by setting all users except the owner to *noaccess*. Depending on the application's context the media was created in, this setting is adjusted. As our scenario outlines, the media was created with MoLES collaboratively. Thus, the media is implicitly shared among users using MoLES together, giving the group members *write* access implicitly. Together, they work on their presentation and engage in discussions with other students. Depending on their location, they share media during discussions, implicitly setting *write* access. When showcasing the presentation in the classroom, the media becomes available to the members of the class, who implicitly gain *read* access. Also, the decision to show the presentation on public displays in school extends the group of users having access to those media. As ALS considers body- and space-related interaction, using and displaying media on displays in a public space implies sharing media *public*ly.

To complete this model, the owner can override implicit settings by setting them explicitly at any time.

In general, our concept of implicitly setting user rights relies on (1) reasonable defaults, which are set upon media creation, (2) the users' interaction with Semantic Media with ALS applications, through which user rights are set implicitly, and (3) explicit user rights, the users may set when necessary. Consequently, using ALS apps in certain *contexts* or in certain *locations* implies certain use cases, which implicate user rights. These are mapped to the user rights model shown in Figure 1. The key issue here is to identify contexts and locations and the effect they are supposed to have, also in the users' affordance when using ALS apps.

All settings for user rights are stored in NEMO in form of semantic annotations (see Figure 1), extending the sets of annotations of Semantic Media. This results in annotation overhead. As every request for Semantic Media is directed to NEMO computing whether a user has access to that media, the overhead produces constant effort. Thus, the user rights settings stored in NEMO through any ALS app have an effect on the accessibility of the media for all the ALS apps connected to NEMO. By this, Semantic Media will also preserve user rights settings when being moved.

IV. CONCLUSION AND OUTLOOK

This article outlines the concept for implicit user rights with regard to the users' interaction with Semantic Media for ALS. Based on a reliable explicit user rights concept, implicit user rights are deduced from the context, in which the media is used, and set through NEMO. The user rights are stored as semantic annotations for each medium and thus are a supplement to explicit user rights. From observations from our research project, we deduct that implicit user rights in many cases replace the necessity of defining explicit user rights. Reciprocally, explicit user rights may overwrite implicit user rights. Although the user rights model presented is sufficient for our research, an extension might be subject for discussion for the use in other contexts. Setting user rights implicitly may presents a security risk, which should be subject of further consideration and could result in new requirements for the user's interaction concepts, e.g., to create awareness when interacting with media. We will continue our observations upon experimental deployment of this concept in NEMO and set up studies to see how well it corresponds with users' needs and expectations.

ACKNOWLEDGMENT

This work is supported by the German Research Foundation (Deutsche Forschungsgemeinschaft, DFG).

REFERENCES

[1] D. Bouck-Standen et al., "Reconstruction and Web-based Editing of 3D Objects from Photo and Video Footage for Ambient Learning Spaces," *Int. J. Adv. Intell. Syst.*, IARIA, vol. 11, pp. 91-104, 2018.

[2] T. Winkler, D. Bouck-standen, M. Ide, A. Ohlei, and M. Herczeg, "InteractiveWall 3.1 - Formal and Non-Formal Learning at School with Web-3.0-based Technology," in *EdMedia*, pp. 1317-1326, 2017.

[3] A. Ohlei, D. Bouck-Standen, T. Winkler, J. Wittmer, and M. Herczeg, "InfoGrid4Museum: A Media Mediation Strategy using Augmented Reality for Museums," in *47. J. d. Ges. f. Inf. eV. (GI)*, pp. 1-8, 2017.

[4] Y. Lu and R. O. Sinnott, "Semantic Security for E-health: A Case Study in Enhanced Access Control," *Proc. - 2015 IEEE 12th Int. Conf. Ubiq. Intell. Comput. 2015*, pp. 407-414, 2015.

[5] P. Ferrara and P. Müller, "Automatic Inference of Access Permissions," *Verif. Model Checking, Abstr. Interpret. - 13th Int. Conf. VMCAI 2012*, no. May, pp. 202-218, 2012.

[6] G. Treu, F. Fuchs, and C. Fuchs, "Implicit Authorization for Social Location Disclosure," *J. Softw.*, vol. 3, no. 1, pp. 18-26, 2008.

Narrative Semantic Media for Contextual Individualization of Ambient Learning Spaces

David Bouck-Standen, Alexander Ohlei, Thomas Winkler, Michael Herczeg
Institute for Multimedia and Interactive Systems
University of Luebeck
Luebeck, Germany
email: [bouck-standen, ohlei, winkler, herczeg]@imis.uni-luebeck.de

Abstract— **Digital conservation and transformation of cultural content and cultural property are constantly increasing. In our research project Ambient Learning Spaces, funded for seven years by the German Research Foundation, we developed a user-centered scenario to individualize and personalize user experience through the use of what we call Narrative Semantic Media. In this context, museum exhibits installations and objects like fossils are digitally augmented into interactive objects for the visitors. Using the visitor's personal mobile devices, our solution embeds the visitor into a flexible, self-adapting narrative structure motivating a self-directed discovery, which creates an Ambient Learning Space inside the museum. The applications we develop running on visitors' mobiles connect to our web- and service-based platform, the Network Environment for Multimedia Objects, a framework that also collects usage and interaction statistics, which are used to personalize the visitors' user experience.**

Keywords-Museum; Narratives; Semantic Media; informal learning;

I. INTRODUCTION

In contemporary approaches, aspects of our cultural property are more and more commemorated by the means of digitalization. Today, any physical object can somehow be transformed into digital media, such as texts, scans, pictures, audio, video footage, or 3D objects [1][2]. These media can be attributed and interconnected within a dynamic information model. For such information models, semantic classification systems for museums, already exist [3].

The transformation from physical to digital space changes the means and expectations, by which exhibits are preserved, augmented, interconnected, and made accessible. Consequently, this will enrich, support and change human perception and receptions of the exhibits. Therefore, the individual visitors play an increasingly important role and should be given the opportunity to critically create their own perspective of the museum and it exhibits [4]. Thus, they become actively and individually involved recipients.

For more than five years, the Institute for Multimedia and Interactive Systems (IMIS) at the University of Luebeck has been developing digitally enriched learning environments for schools and museums in the research project *Ambient Learning Spaces (ALS)* [5], supported by the German Research Foundation (Deutsche Forschungsgemeinschaft, DFG). In context of ALS, body- and space-related human-computer interaction, as well as the concept of cross-device interaction (XDI), builds the conceptual foundations. For this purpose, a family of connected learning applications has been developed as the frontend systems for visitors and curators. Providing these applications, we focus on self-directed learning with interactive media [6].

In the backend, the *Network Environment for Multimedia Objects (NEMO)* is the platform for all ALS applications [7]. NEMO stores all media created by learners in a specific semantic model, which supports the use by ALS applications and reflects the respective applications context. The ALS applications' run on mobile or stationary devices and access NEMO as a contextualized repository. Media created in NEMO can be interrelated inside NEMO and will be presented to the visitors as augmentations of the physical exhibition on different interconnected digital devices like smartphones, tablets or large stationary multi-touch screens [6]. For the visitors, this creates an individual multi-perspective experience of the exhibitions based on personal dynamic narrative paths.

Narratives have always played a central role in museum didactics. It has been first of all "stories" that, physically, verbally, and later also technologically enhanced, guided the visitors through the exhibits [8] on a more or less linear path. In our project, we chose two very different museums as project partners. As testbeds for field studies, we are working closely together with the *Günter Grass-House* (GGH), dedicated to late Nobel Prize winner Günter Grass, and the *Museum of Nature and Environment* (MNU). Both museums belong to the foundation "Die Lübecker Museen" located in the Hanseatic City of Luebeck in Germany and require applications, which allow a personal view of the exhibition in a multi-perspective way [9]. Thus, any narrations may no longer consist of the same and expected, but rather the unexpected, alternatives, or evolutions. Museum visitors shall be able to actively construct and deconstruct their individual and contextualized stories through their interaction in the physical museum space.

In our prototypical implementation, we use web-based systems to provide museum visitors with this personal narrative experience. In general, regarding the user interfaces, these systems are browser-based, with device-specific extensions. Our solution allows museum curators, as well as other museum professionals, to construct the digital augmentation of their physical exhibitions. Inside the museum, visitors will use InfoGrid as one of our ALS applications on their own mobile devices they are already accustomed to [10]. InfoGrid [11] augments physical objects of the exhibition with digital media, e.g., by augmented reality (AR) and guides visitors through the exhibition by what we call *Narrative Semantic Media (NSM)*. Museum visitors will also use the InteractiveWall [6], a software system for large multi-touch displays containing applications, e.g., displaying media from ALS.

Our solution utilizes a concept of semantically enriching media in a particular way. All media handled through NEMO are modeled as NSM. NSM consists of media, such as text, still images, audio, video footage, or 3D objects, which NEMO stores as binary data. In NEMO, they are enriched by semantic attributions. The semantic attributions consist of a basic set of annotations, which are used internally by NEMO, as well as an extended set of annotations, which are specific for the context the media object is dedicated to. This context is projected into the semantic model for, e.g., the GGH or MNU, which we focus upon in more detail below. The semantic annotations describe the content and define its place in the semantic model. Although pedagogical and user-related implications with regard to narrations exist, in this contribution we focus on the technical solution.

In this contribution, in Section 2, we regard related work. In Section 3, we illustrate our research scenario. In Section 4, we outline the system architecture. In Section 5, we describe the use of narrative semantic media in museums, and discuss results from the pre-evaluation and validation in Section 6. In Section 7, we present the conclusion and outlook.

II. Related Work

With our work, we focus on dynamic individual narratives, as well as the question of the possible influence of technology on the self-conception and the role of museums. Visitors entering a museum will come with substantial knowledge with potential references from and to the museum. However, depending on the individual, the type of museum, the exhibition, the level of visitor's knowledge will range from complete novice to expert. Therefore, the most satisfying exhibitions for visitors will be those that resonate with their own experience and provide information in ways that confirm and enrich their personal view of the world. Conflicting perceptions induce personal critical reflections that may change knowledge and opinions. Visitors want museum visits to be inspirational and uplifting, emotionally developmental in some way, but also to be picked up with their existing points of view [12].

By digitally individualizing narrative structures inside the museum, new productive relationships between the curator and visitors may be established, especially if in the process of individualization a visitors profile is used to automatically provide the visitors with media potentially referring to the visitors' knowledge and expectations. At the same time, however, the massive use of multimedia systems in museums are seen critical because of the potential decrease in the role and power of physical objects in museums [13].

With our work, we focus on using the outcome of the process of digitalization of museum content. According to Hyvönen [14], publishing linked Cultural Heritage collections creates a major challenge for interoperability. With the applications developed in ALS, we present a solution to make use of digital media collections to extend physical exhibitions with digital content. For museum visitors, we provide applications, which enable visitors to interact with the digitally augmented exhibition in an individual multi-perspective experience. Enhancing a physical exhibition by the means of AR has already been accomplished in various related works. As in 2008, Miyashita et al. already presented an AR museum guide [15], this approach is not new, but the approach presented in this article embeds the visitors into a flexible, self-adapting narrative structure and thereby performs the augmentation. This also distinguishes our work from digital story-telling, like the approaches of Vayanou et al. [4], Antoniou et al. [16], Boy et al. [17], or van Dijk et al. [18]. These approaches, e.g., have in common to require data to be created or especially prepared for use in narrations. Although certain elements of our solution have been inspired by digital storytelling, our approach automatically generates a narration from semantically annotated media with only minimal annotations to the artefacts.

Using NSM through NEMO, the visitor's experience is individualized, making use of a knowledge management platform architecture, as Dragoni et al. outlined [19]. Distinguishing our solution from the approach of Blumenstein et al. [20], the visitors own mobile devices will be integrated seamlessly into the museum's technological ecology while accessing NEMO, interconnecting to other technical museum systems. We also use the mobile devices to identify the museum visitors and thus, personalize their museum experience. Other ways of identification will be possible.

III. Scenario

Although this work presents a backend solution, we examine the following scenario of human-computer interaction to illustrate the backend work:

Maureen O'Grady enters the Günter Grass-House through the shop entrance and buys a ticket. Next to the cash register, she notices a poster advertising the mobile app InfoGrid. Becoming even more curious, she downloads and installs the app on her smartphone. Upon startup, InfoGrid is asking some questions related to personal data and preferences. Maureen answers them, e.g., by selecting "literature", "graphic art" and "sculpting" from a list of semantic tags suggestions. She also specifies that she has an hour time left for her visit to the GGH, She finishes this phase after just a minute and finally starts the tour "Günter Grass: My Century" within InfoGrid.

Maureen leaves the shop while InfoGrid displays a hint indicating that she can always use the digital tour of "My Century" wherever she notices a depicted graphical symbol inside the exhibition. For now, she puts her smartphone away and continues through the courtyard and from there through a glass door into the foyer. On the right wall of the foyer, a large shelf with numerous items is attached, the "Cosmos Grass". This is a collection of various items Günter Grass made or collected by himself. A sign says, "Please touch". At the table in the middle of the room, the "Curator's Table", her smartphone suddenly vibrates. She pulls it out of her pocket and follows the instructions on the screen of InfoGrid. Holding her smartphone in a way that the camera captures the

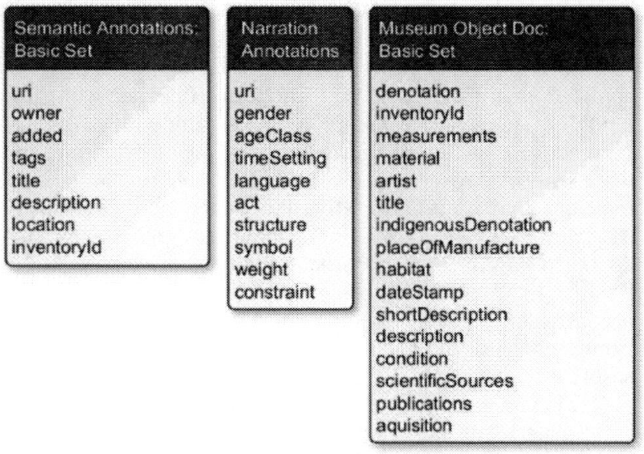

Figure 1. Overview of annotation sets for NSM.

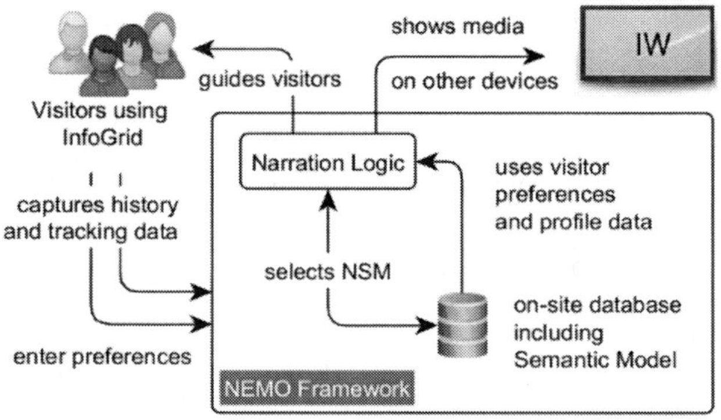

Figure 2. Based on visitors' preferences, history and tracking data, NEMO selects NSM in order to guide visitors with InfoGrid and additionally controls media displayed other devices in the museums.

Curator's Table through her smartphone, she can see the sculpture of "Seven Birds" standing right up on the table like being real. Intuitively she tries to grab the virtual sculpture. Walking around the table, she looks at the high resolution virtual sculpture from all sides. Then, she puts the smartphone back into her pocket and, after taking a closer look at the "Cosmos Grass" and despite another symbol for "My Century", for which she has not taken out her smartphone again, she continues into the next room.

On the wall of the next room Maureen slowly passes the InteractiveWall. She notices how the wall's display suddenly changes, just as she approaches the screen. Instead of the general information displayed before, a media gallery with images and videos gets visible. The gallery is showing Günter Grass with some of his tools in his sculpture workshop. Maureen is thrilled to see how Günter Grass actually created the sculpture "Seven Birds" she had previously viewed from all sides as a 3D object in InfoGrid. She continues her way through the GGH smiling, as sculpturing is a special interest of her as she told InfoGrid earlier.

Finally, Maureen makes her way to the exit. From a distance, she sees how another visitor passes the InteractiveWall screen and the content changes again. Curious, she comes closer and sees that completely different media have been displayed for the other visitor. He notices Maureen and they start a conversation about this phenomenon and their personal interests.

IV. System Architecture

Museum visitors can use InfoGrid, described in more detail by Ohlei et at. [11], to experience information tours at specific locations with their own mobile devices. These locations can be inside a museum building as well as in urban or rural areas. With the help of web-based ALS software, curators can create or edit these information tours. The visitors can download InfoGrid inside the museum through regular Android or iOS app stores. Once installed and started, InfoGrid connects to an instance of NEMO on-site and displays NSM. NEMO individually selects these NSM for each visitor, e.g., from personal preferences the visitors enter on startup of InfoGrid.

InteractiveWall, as introduced by Winkler et al. [6] is a web-based application, which runs on systems equipped with 32"-90" multi-touch screens. Each installation of the InteractiveWall is equipped with a Bluetooth LowEnergy beacon, which allows InfoGrid to detect the visitor's position in the physical exhibition. This supports the individualized display of content on the InteractiveWall through NEMO, e.g., when a visitor is in front of the installation.

NEMO is a web-based framework for ALS, described by Bouck-Standen et al. in more detail [21]. The NEMO API provides access for applications, such as InfoGrid, interacting through Web Services in an authenticated context over a secure connection. The NEMO framework decides which media is presented to the visitor depending on multiple algorithms.

V. Using Narrative Semantic Media in Museums

The scenario illustrates how InfoGrid guides the user through the exhibition. In the following, we describe how with the use of NSM, our backend solution provides the necessary media for display in InfoGrid.

At first, Figure 1 shows the annotation sets and included tags required for our setup. With the annotations from the basic set, the media is described with a subset of annotations of the Ontology for Media Resources [22], which is sufficient for our purposes. These are compatible with the basic Museum Object Documentation, which Dresch and Mainberger [22] examine as a link between the real museum collection and the digital reflection of a museum. This is also utilized in museum documentation systems [3]. These implement the relationship between the systems presented in this article to museum systems already established. In addition to semantic annotations, narrative annotations extend media, as depicted in Figure 1. These are required by NEMO in order to place NSMs on narrative paths.

In the following example, a narrative path has a fixed length l and consequently consists of l NSMs, which are selected by the NEMO applying to a story model. According to the visitor's settings, e.g., depending on the time she or he can spend inside the museum, l is varied. The visitors set the

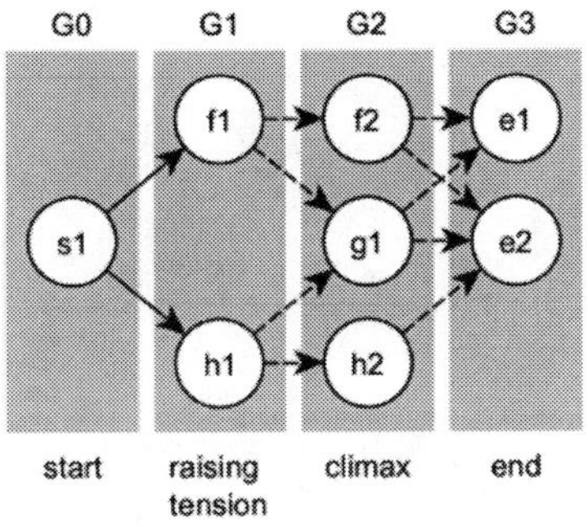

Figure 3. Exemplary narrative paths of length l=4 generated from a total of 8 NSM.

length *l* on startup of InfoGrid, when the application asks, how much time they would like to spend in the exhibition.

As

Figure 2 illustrates, NEMO selects Semantic Media based on the on-site data, semantic model and visitors input, interaction and usage history. By selecting a sequence of media and providing them to InfoGrid as a narrative path, the mobile app is able to guide the visitors by displaying the contents of the NSM to them according to their location inside the exhibition. During usage, InfoGrid continuously submits data about the visitor's interactions in a way that NEMO is able adjust the selection of media during the visitor's stay. Consequently, visitors do not necessarily follow a preset or given path through the museum, but are able to explore depending on their interest or intent raised by this digitally enriched exhibition. This can result in a possible change of the narrative path.

For the scenarios presented in this article, we focus on a basic story model based on arc of suspense [23]. This story model is characterized by an introductory start, a part of raising tension followed by a climax, an optional phase of falling tension (outro) and an end, as illustrated in Figure 3. Technically, this requires the NSM to be assigned into these phases to determine the position of an NSM on a narrative path. In this implementation, the annotation *act* (see Figure 1) positions an NSM with the values of *start, end, predecessor, successor*, and *context* at the corresponding position of the narrative path. There are no special annotations for raising tension or climax, to allow those phases to consist of an unlimited number of NSM. Thus, the suspense of a narration is implicitly modeled by the positions of NSM through the *act* annotation. Other annotations shown in Figure 1, such as e.g., *ageClass*, *gender*, *language*, or *constraint* are used in order to select NSMs according to the visitor's profile.

The creation of a new narrative for a tour through an exhibition begins with the selection of media from the pool of media the museum has already semantically annotated. At first, for the start of the tour, an NSM has to be selected, which Figure 3 illustrates as *s1*. In the example, the tour should end with *e1* or *e2*. For this, these media are annotated with the corresponding annotation. Consequently, Figure 3 shows, that a visitor can reach two possible ends.

Having defined the start and end or a tour, all NSMs have to be identified the tour may consist from. In this case, there are *5* NSMs from the pool of media, which are suitable for the exhibition at hand. In this example, the media named *f1* and *h1* have the same *structure* annotation, and are both as successors of *s1*. The NSM named *f2*, *g1* and *h2* are also labeled with same *structure* annotation, which is different from *f1* and *h1*. As a result, *f1* and *h1*, as well as *f2*, *g1*, and *h2* are grouped in two separate groups, *G1* and *G2* as Figure 3 shows. It can be observed from Figure 3, that starting from *s1* the narrative path branches out into *f1* or *h1*. As those NSM to follow *s1* on the narrative path are set manually, Figure 3 shows the transition as undashed arrows. For visitors starting at *s1*, NEMO determines which path to take. This depends on the semantic annotations mentioned above in context with the users' profile and history. If, e.g., *f1* and *h1* with a distinct *ageClass* annotation set were to distinguish different age classes of visitors, children would follow the path *p1=s1, f1*, whereas adults *p2 = s1, h1*, for example. Moving from *G1*, which contains NSM to raise tension, to the climax, which all NSM selected for *G2* represent, the transitions Figure 3 shows as dashed arrows are automatically determined by NEMO. For this, at first, *G2* is automatically aligned in sequence after *G1* due to *timeSetting* annotations of the NSM. As no other groups exist, *G3* containing both ends *e1* and *e2* follows sequentially after *G2*, and the transitions Figure 3 are calculated equivalently. Afterwards, for each path leading into *G2*, from semantic and narrative annotations, such as tags, location, or constraints, NEMO determines, which NSM follows on the paths *p1* or *p2*. In the exemplary abstract case shown in Figure 3, *p1*, e.g., may be extended by both *f2* and *g1*, as *g1* can be served both age classes mentioned above. This results in *7* possible narrative paths of length *l=4*.

At any point, other additional transitions could be defined manually by adjusting the *successor* or *predecessor* annotation in *act* of the corresponding NSM. This provides the means to directly adjust narrative paths.

The length *l>=4* can be chosen by the visitor. However, a predefined narrative path with *l=10* may be shortcut by visitors who lack time for completing the tour in full. In this case, the *act* and *structure* annotations make sure that the most important NSMs are included in the narrative path. In addition, NEMO is not limited to the story model introduced here.

The narrative path calculated by NEMO also depends on the visitor's physical location inside the exhibition. We detect the visitor's location using InfoGrid and Bluetooth LowEnergy beacons, which are positioned at key points of the installation, such as room entrances. If a visitor is near such a point of interest, NEMO triggers recalculation of the NSM distributed to the instance of InfoGrid of the respective user.

In Figure 4, we regard the world model of the GGH, which abstracts the museum, its exhibits, objects and research into digital space. This brings the late artist and writer Günter Grass into context with, e.g., persons he knew, was influenced by, and influenced upon, institutions, elements from his biography, elements from his world picture, and elements from his works. For complexity reasons, Figure 4 shows an overview of the model with focus on the sub-model used for

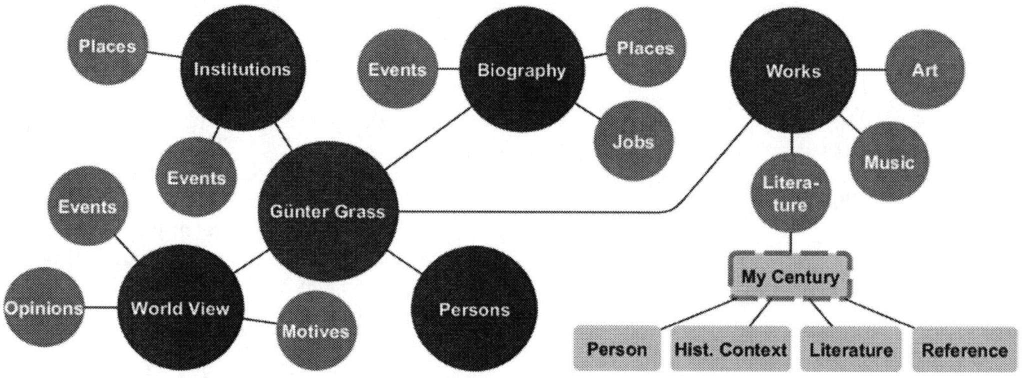

Figure 4. The museum model of the GGH from an overview perspective at the current state of research.
For the scenario presented in this contribution set in the GGH, we focus on the use of NSM in the "My Century" part of the model.

"My Century" in our scenario. For the Museum of Nature and Environment another model exists accordingly.

Grass "My Century" casts a retrospective review of the 20th century in a hundred stories and is narrated from the point of view of different people from all areas of German society, from an assembly line worker to a professor. Each story is accompanied by an artwork created by Grass and features crosslinks to persons, historic events, and other works of Grass, which are mostly hidden in textual hints or, e.g., allusions.

As the above process of creating a narrative path with the use of NSM outlines, a user is required to define NSM as *start* and *end* narratives. However, the process of selecting narratives between starting and ending narratives can be automated. We achieve this by using the semantic model created for the particular context the media is selected in. For example, for a certain year from "My Century", according to the semantic model (see Figure 4), NSM annotated from that year correlates to other NSM, not necessarily dated to the same year, and possibly annotated corresponding to the visitor's interest. Thus, the process of media selection for any other than *starting* or *ending* NSM, NEMO automates the selection, leaving annotations such as *ageClass*, *gender*, *language*, or *constraint* for manual retouch only.

VI. Pre-Evaluation and Validation

We have carried out a pre-evaluation of our implementation under laboratory conditions with 5 subjects using a total of 50 NSM with both 3 starting and ending narratives. The evaluation was conducted using the arc of suspense as the story mode. In our tests, l was defined as $l \in \{3, 7\}$ and we configured NEMO to automatically create two tours per subject. The tours were displayed inside a web-browser and we collected the user's feedback according to the *Think-Aloud* method [24]. In all cases, the subjects experienced both tours and gave positive feedback. The NSMs were assembled to reasonable narrative paths that the subjects were able to comprehend. Under laboratory conditions, with 50 NSM and both 5 starting and ending narratives, NEMO was able to construct narrative paths of all lengths l with $l \in \{3, 10\}$. The test show that, apart from the duration a visitor is willing to spend on a narration, the upper boundary of l depends on the number of semantic connections between the NSMs. However, we have not defined a measure for the complexity of semantic correlations between NSMs to be able to make a general statement with regard to the capabilities of NEMO being able to always create a narrative path to the visitors' desire of length l. Although NEMO was able to generate the narrative paths in our test settings from only 50 NSMs, there may exist scenarios, where NSMs may not be connected to each other. A solution to this challenge may present itself by selecting a story model with a looser coupling between the NSM. The postmodern story model [25] allows more combinations between NSMs, as the model does not follow an arc of suspense. We have also implemented and evaluated this story model with the subjects and analogous to the setting described above. However, the subjects clearly indicated that they did not understand the concept of the postmodern story model and experienced the narrative paths generated by NEMO from the postmodern story model as "disordered" or "incomprehensible". Museum professionals from our project partners confirmed this observation.

The pre-evaluation shows, that NEMO creates individual narrative paths that the subjects are able to understand and follow. This influences the perceived quality of the narration. The pre-evaluation indicates that the parameters of perceived quality and length of narrative paths generated from NSM have to be considered together. It also suggests that (a) the quality of narrative paths perceived by visitors depends on both quality and quantity of NSMs available, on which the narrative paths are based and from which they are generated by NEMO, as well as (b) the complexity of the story model, which is also dependent on the type of audience.

For future evaluations, from the qualitative feedback and from the experience of our museum project partners, we derive that a questionnaire should focus on user experience and be accompanied by questions with regard to the visitors understanding of the story model, the visitors' perception of the narrative path in total, and the perceived quality of the NSMs selected by NEMO as to their motivation of self-directed discovery.

VII. Conclusion and Outlook

In this contribution, we illustrate a system concept to provide an individualized museum experience. The goal is to augment physical museum exhibits with digital media in a personalized, flexible, and self-adapting narrative structure motivating self-directed discovery. Based on the scenario, we describe the use of technology to design an individual experience for visitors. Unlike other approaches of storytelling, NEMO generates narrative paths from Narrative Semantic Media, presented in this contribution. Thus, museum narrations consist of media as digital overlays and extensions to physical exhibits. Narrative Semantic Media is one focus of our research and a main concept throughout ALS. In NEMO, we use semantic annotations to classify and interrelate plain media within semantic models, and algorithms accessing the semantic models and the users' personal information, e.g., on interests or interaction history, in order to select semantic media for narratives. These media are enhanced for their use on dynamic narrative paths through narrative annotations, forming what we call Narrative Semantic Media. Having selected NSM for a narrative path for a thereby personalized user experience, NEMO depends on ALS applications, such as InfoGrid and the InteractiveWall. These frontend applications allow the users, in our scenario the museum visitors, to interact with NSM.

In our future work, we will also study the visitor's experience on-site with our project partners quantitatively and in more detail. Through the semantic relation of the stored data, visitors can discover new relationships between the museum objects or even urban structures outside the museum. Next to the visitors' experience, we will study the curator's experiences while using ALS systems. Furthermore, we will evaluate how curators build and how visitors use dynamic narrative paths throughout the museum. As InfoGrid and the InteractiveWall are part of the family of ALS applications, we plan to evaluate their interconnection with other applications, also in school context. In this context, the interconnection between multiple distributed instances of NEMO will be subject to further research and development.

Acknowledgment

We develop NEMO, InfoGrid, and other ALS-applications in the research project Ambient Learning Spaces funded since 2008 by the German Research Foundation (Deutsche Forschungsgemeinschaft, DFG).

References

[1] D. Bouck-Standen, A. Ohlei, V. Daibert, T. Winkler, and M. Herczeg, "NEMO Converter 3D : Reconstruction of 3D Objects from Photo and Video Footage for Ambient Learning Spaces," in *AMBIENT 2017; The Seventh Int. Conf. on Ambient Comp., Applications, Services and Technologies*, IARIA, *Best Paper Award*, pp. 6-12, 2017.

[2] R. Li, T. Luo, and H. Zha, "3D digitization and its applications in cultural heritage," in *Lecture Notes in Computer Science (including subseries Lecture Notes in Artificial Intelligence and Lecture Notes in Bioinformatics)*, vol. 6436 LNCS, M. Ioannides, D. Fellner, A. Georgopoulos, and D. G. Hadjimitsis, Eds. Berlin, Heidelberg: Springer Berlin Heidelberg, pp. 381-388, 2010.

[3] "Europeana," 2018. [retrieved: August, 2018]. Available: europeana.eu.

[4] M. Vayanou, et al., "Authoring Personalized Interactive Museum Stories," ICIDS 2014: Interactive Storytelling, pp. 37-48, 2014.

[5] T. Winkler, F. Scharf, C. Hahn, and M. Herczeg, "Ambient learning spaces," *Educ. a Technol. world Commun. Curr. Emerg. Res. Technol. efforts*, pp. 56-67, 2011.

[6] T. Winkler, D. Bouck-standen, M. Ide, A. Ohlei, and M. Herczeg, "InteractiveWall 3.1 - Formal and Non - Formal Learning at School with Web - 3.0 - based Technology in Front of Large Multi-touch Screens," EdMedia, *Outst. Paper Award*, Washington, DC, pp. 1317-1326, 2017.

[7] D. Bouck-Standen, "Construction of an API connecting the Network Environment for Multimedia Objects with Ambient Learning Spaces," University of Luebeck, 2016; DOI: 10.13140/RG.2.2.12155.00804.

[8] B. MacIntyre, J. D. Bolter, E. Moreno, and B. Hannigan, "Augmented reality as a new media experience," *Proc. - IEEE ACM Int. Symp. Augment. Reality, ISAR 2001*, pp. 197-206, 2001.

[9] L. Pujol, et al., "Personalizing interactive digital storytelling in archaeological museums: the CHESS project," *Archaeol. Digit. Era. Pap. from 40th Annu. Conf. Comput. Appl. Quant. Methods Archaeol. (CAA)*, pp. 77-90, 2012.

[10] B. Al Takrouri, K. Detken, and C. Martinez, "Mobile Holstentour: Contextualized Multimedia Museum Guide," *MoMM 2008*, pp. 460–463, 2008.

[11] A. Ohlei, D. Bouck-Standen, T. Winkler, J. Wittmer, and M. Herczeg, " InfoGrid4Museum: A Media Mediation Strategy using Augmented Reality for Museums," in *47. Jahrestagung der Ges. für Informatik e.V. (GI)*, pp. 1-8, 2017.

[12] Z. D. Doering and A. J. Pekarik, "Questioning the Entrance Narrative," *Source J. Museum Educ.*, vol. 21, no. 3, pp. 20-23, 1996.

[13] A. Davis and K. Smeds, *Visiting the Visitor: An Enquiry Into the Visitor Business in Museums*, Edition Mu. Transcript Verlag, 2016.

[14] E. Hyvönen, "Cultural Heritage Linked Data on the Semantic Web: Three Case Studies Using the Sampo Model," in *Invited talk, Proc. of: VIII Encounter of Documentation Centres of Contemporary Art: Open Linked Data and Integral Management of Information in Cultural Centres*, 2016.

[15] T. Miyashita, p. Meier, and T. Tachikawa, "An Augmented Reality Museum Guide," *Proc. - 7th IEEE Int. Symp. Mix. Augment. Real., ISMAR 2008*, pp. 103-106, 2008.

[16] A. Antoniou *et al.*, "Capturing the visitor profile for a personalized mobile museum experience: An indirect approach," *CEUR Workshop Proc.*, vol. 1618, el. ed., 2016.

[17] J. Boy, F. Detienne, and J.-D. Fekete, "Storytelling in Information Visualizations," *Proc. 33rd Annu. ACM Conf. Hum. Factors Comput. Syst. - CHI '15*, pp. 1449-1458, 2015.

[18] B. van Dijk, A. Lingnau, G. Vissers, and H. Kockelkorn, "Playful User Interfaces," *Play. User Interfaces, Gaming Media Soc. Eff.*, no. February, pp. 185-208, 2014.

[19] M. Dragoni, S. Tonelli, G. Moretti, and F. B. Kessler, "5 A Knowledge Management Architecture for Digital Cultural Heritage," *ACM J. Comput. Cult. Herit. Artic.*, vol. 10, no. 18, pp. 1-18, 2017.

[20] K. Blumenstein, M. Kaltenbrunner, M. Seidl, L. Breban, N. Thür, and W. Aigner, "Bringing Your Own Device into Multi-device Ecologies," *Proc. Interact. Surfaces Spaces ZZZ - ISS '17*, pp. 306-311, 2017.

[21] D. Bouck-Standen, et al., "Reconstruction and Web-based Editing of 3D Objects from Photo and Video Footage for Ambient Learning Spaces," *Int. J. Adv. Intell. Syst.*, vol. 11, IARIA, pp. 91-104, 2018.

[22] V. Huck-Fries, F. Wiegand, K. Klinker, M. Wiesche, and H. Krcmar, " Data glasses during maintenance: Evaluation of various input modalities by service technicians," *Inform. 2017*, pp. 585-596, 2017.

[23] B. O'Neill, "Toward a computational model of affective responses to stories for augmenting narrative generation," in *Lecture Notes in Computer Science (including subseries Lecture Notes in Artificial Intelligence and Lecture Notes in Bioinformatics)*, vol. 6975 LNCS, no. PART 2, Memphis, TN: Springer-Verlag, pp. 256-263, 2011.

[24] J. Nielsen, *Usability Engineering*. San Diego, CA: Kaufmann, 1994.

[25] M. Currie, *Postmodern Narrative Theory*. Palgrave Macmillan, 1998.

Preprocessing Data for Machine-Learning Algorithms to Provide User Guidance in Special Purpose Machines

Sascha Lang*, Valentin Plenk†
Institute of Information Systems at Hof University
Hof, Germany
Email: *sascha.lang@iisys.de, †valentin.plenk@iisys.de

***Abstract*—In our previous work, we proposed a system which makes complex production machines more user-friendly by giving the recommendations to the operator. So, we assist the user working with a complex production machine. The recommendations are presented like: "In the last 10 occurrences of this event the operators performed the following keystrokes". While working on the project, we had problems with retrieving the correct recommendations from our knowledge base. Meanwhile, we gathered more data from our project partners. Now, we dive deeper into this data in order to improve our solutions. This work describes methods to preprocess the data. This preprocessing should help us building up the knowledge base. To achieve this automatically, we do not want to know much about the machine and the production process itself.**

***Keywords*–machine-learning; human machine interfaces; special-purpose machines; production machines; case based reasoning**

I. INTRODUCTION

In [1], we used machine learning algorithms to generate recommendations like "in the last 10 occurrences of this event the operators performed the following keystrokes" for operators of complex production machines. Figure 1 shows the basic structure of a plastic extruder. Figure 2 draws the basic structure of our system. The left side represents the machine, in our case, the extruder. The right side shows the structure of our system. Its purpose is to build up a knowledge base from operator interactions which were performed in the past. Since, we started our project, the surroundings stayed the same, as discussed in [2].

Our system is faced with two challenges: First is retrieving the recommendation that is most suitable for the current state of the production machine. Second is building up a knowledge base by extracting the operator interactions from data gathered from the production machine.

In our previous paper [3], we evaluated our existing algorithms, which were created to retrieve recommendations from our knowledge base. But because, we were not satisfied with the results, we decided to take one step back. Instead of improving our retrieval algorithms, we now focus on the way we create the knowledge base from the data, we got from our project partners. We developed some improvement steps, and we will check how they influence our retrieval results .

A similar work is done by N. Ben Rabah et. al. [4]. They also create their knowledge base automatically. In contrast to our project they use two knowledge bases: One for normal states and one for failure states. We on the other hand use only one knowledge base. This solely stores the situations, in which the machine needs attention. Another difference is that their data only consists of binary values. We are faced with all kinds of numerical data. Beside bit values, we also deal with floating point numbers of different ranges.

Some other approaches, like the "APPsist" system [5] or the system described by Cen Nan et.al. [6] have a manually build up knowledge base. We on the other hand focus on creating the knowledge base autonomous. So, we are faced with the challenge to recognize important subsets of our gathered data automatically, which means without the help of a process expert.

Before providing a possible solution, we have to define some datasets. Therefore, we created three data sets from different machines. All machines are plastic extruders. The ones used for Dataset 5 and Dataset 6 contain the same control system. The machine for Dataset 7 uses a different type of

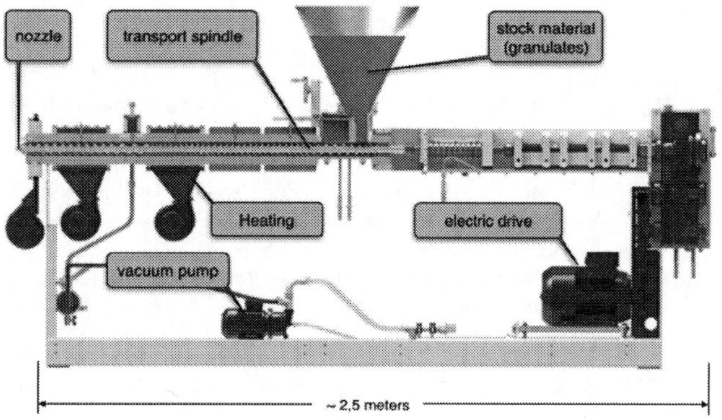

Figure 1. Plastics extruder
(courtesy: Hans Weber Maschinenfabrik, Kronach, Germany)

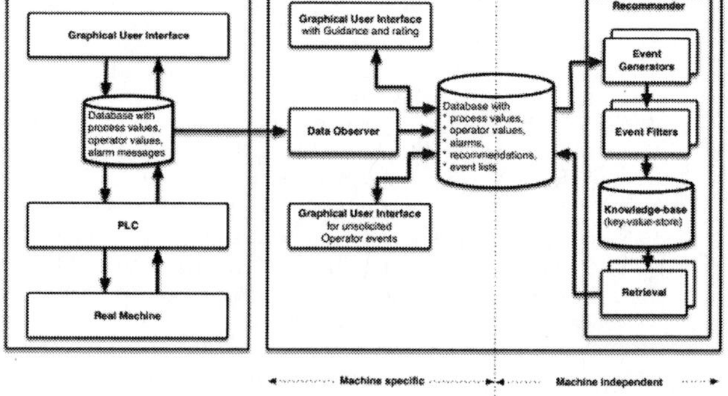

Figure 2. System architecture

control system which contains no historical data. The period how long each dataset lasts, was registered. We also logged the time for which the machine was productive during that period. These information can be seen in Table I. The biggest difference we found between the datasets relates to the amount of user operations we found. Dataset 7 has far the least amount of operations but nearly the longest runtime.

TABLE I. DESCRIPTION OF THE TEST DATASETSS

	Dataset		
	5	6	7
Start	17-11-05	17-11-05	17-11-26
End	18-03-05	18-04-23	18-04-23
Runtime	$\approx 758h$	$\approx 1664h$	$\approx 1645h$
Events	1159	537	147

The data, we get from our productions machine are stored in a database table, an example is shown in Table II. Each column, contains the value either for a process value or a operator value. We get this data every 10 seconds. The tables for Dataset 6 and 7 are similar, but have a different amount of columns. In our data, we have numerical values as for temperature and rotational speed as well as percentage and bit values.

The next section describes the problems, we discovered while working on our prior approach. Section III describes how, we process the data from the machine to create the values for the knowledge base. In Section IV, we describe our methods to get better data for the keys. The last Section V summarizes the results and gives a short description of the problems, we have to solve in the future.

II. PROBLEMS WITH PRIOR APPROACH

In our prior approach, the trigger point for an entry in our knowledge base was the beginning of an alarm situation. For this, we needed an additional table containing the start and end times of every alarm. For every alarm, we generated a fingerprint and an operator sequence. This pair represents a key (fingerprint) value (operator sequence) entry in the knowledge base. If a new alarm is raised, we will create the respective fingerprint for it. Now, we search for the most similar key inside the knowledge base. The value belonging to this key is then used to present a recommendation to the operator.

A. Problems with alarms

One problem, we had with our alarm based approach is that, we have to deal with simultaneous alarms. For example, if a temperature alarm for a specific zone is raised, it would also raise general temperature alarm additonaly.

In our data, we discovered three types of simultaneous alarms. The first type has the exact same start and end timestamp, so therefore it lasts for the exact same time period and is simply treated as one alarm. The second group are nested alarms. In this case one alarm is raised while another alarm is active. And it is finished before the first alarm is finished. The last group are so called staggered alarms. In this case a second alarm is raise while another one is still active. But in contrast to the nested alarms it is finished after the first alarm.

Alarms lasting the exact same time period, are treated one alarm. Now let s be the start of an alarm and e its end and

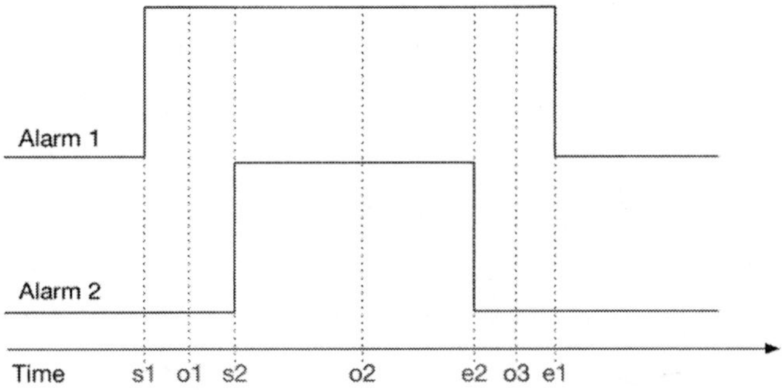

Figure 3. Nested Alarms

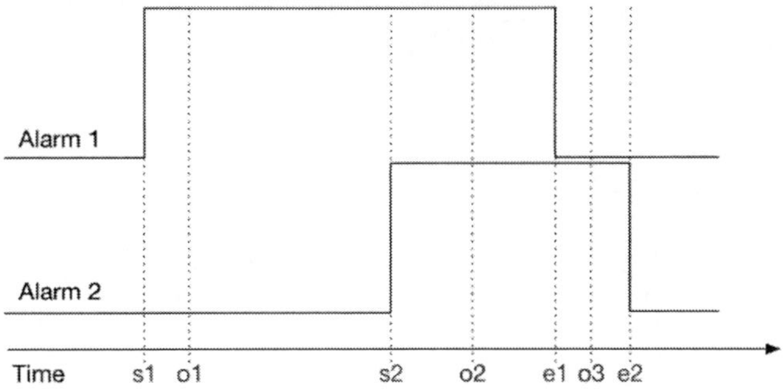

Figure 4. Staggered Alarms

be o some operator events occured during the alarm situation. According to Figure 3 and Figure 4, we were faced with the following situations:

- If $o1$ and $o3$ occur, we can say that the ne alarm should last from $s1$ to $s2$.
- If only $o2$ occurs or $o2$ and $o3$ occur, it is not sure if $s1$ or s2 should be the start of the resulting alarm.
- If only $o1$ occurs or $o1$ and $o2$ occur, it is not sure if $e1$ or $e2$ should be the end of the resulting alarm.
- If only $o2$ occurs, it is not sure which start and end point would be correct for the remaining alarm.

B. User interaction beside of alarms

While analyzing the data more intense, we discovered another problem. We noticed that, we had some user interactions beside the alarm situations. So if, we only consider the situations, with an alarm being raised, we will not consider all situations, in which it will be necessary to change settings of the machines. But because, we want to support the user during the whole time, we have to consider all operator events not only the ones raised during an alarm.

C. Problem with similarity of fingerprints

By solving these problems, we created a knowledge base for every dataset. But now, we noticed some issues with our fingerprints. We expected that entries with the same value (operator-sequence) would have similar keys. But this expectation was not satisfied, we have to figure out how, we can solve this problem.

TABLE II. EXAMPLE DATA FROM DATASET 5

Time	Extruder.Temp0.PV	Extruder.Temp0.OP	Extruder.Temp0.SP	Meltpump.Out.Val	Production_Mode
17-11-05 00:00:06	200.0	34.0	200.0	26.400	1
17-11-05 00:00:16	200.1	34.0	200.0	26.400	1
17-11-05 00:00:26	200.1	30.2	200.0	26.400	1
17-11-05 00:00:36	200.1	33.3	200.0	26.400	1

TABLE III. LIST OF DELAYS BETWEEN EVENTS IN DATA SET 5

Delay(s)	Count	Delay(s)	Count
$d \leq 10$	823	$10 < d \leq 60$	89
$60 < d \leq 160$	17	$160 < d \leq 300$	17
$300 < d \leq 600$	37	$600 < d \leq 1200$	36
$1200 < d \leq 2400$	18	$2400 < d \leq 3600$	8
$3600 < d \leq 7200$	26	$7200 < d \leq 10800$	18
$10800 < d \leq 14400$	6	$14400 < d \leq 18000$	3
$18000 < d \leq 36000$	17	$36000 < d$	43

TABLE IV. LIST OF DELAYS BETWEEN EVENTS IN DATA SET 6

Delay(s)	Count	Delay(s)	Count
$d \leq 10$	128	$10 < d \leq 60$	37
$60 < d \leq 160$	13	$160 < d \leq 300$	8
$300 < d \leq 600$	12	$600 < d \leq 1200$	13
$1200 < d \leq 2400$	24	$2400 < d \leq 3600$	18
$3600 < d \leq 7200$	25	$7200 < d \leq 10800$	14
$10800 < d \leq 14400$	5	$14400 < d \leq 18000$	7
$18000 < d \leq 36000$	29	$36000 < d$	72

III. Generation of Operator Sequences

To solve the problems described in Section II, we decided to not consider alarms from now on. Because, we are able to extract operator events from the machine data, we will use this events as a trigger point. Around these trigger points, we will generate some kind of an own "alarm".

To create operator sequences to the respective the events, we have to decide how long the pause between two events should be before they are considered as different sequences.

A. Using the Hauloff Speed

Our first idea to determine this pause is, to evaluate, how much time an operator needs to make changes to the machine. So, we interviewed some machine operators. These conversations revealed that the time depends on the product. How fast a product is created depends on the so called Haul Off Speed. Together with an information of the extruder length, we can estimate the time of the intermission.

Unfortunately none of our test systems provides the Haul Off Speed. So, we talked to a process expert to give us an estimation for a rough maximum and minimum value. In our case the main products are pipes, so the expert mentioned that the haul off speeds will be between $1\frac{m}{min}$ and $50\frac{m}{min}$. If, we have an extruder being, e.g., 35 m long, the resulting period is between $\approx 45s$ and $\approx 2100s \approx 35min$.

B. Using a constant ratio

For this method, we create a list of all pauses between one event and its subsequent event. This list is ordered from small to long delays. We now assume that, e.g., the biggest 20 percent of this delays are pauses between two sequences. The event counts for Dataset 5 is shown in Table III, for Dataset 6 in Table IV, for Dataset 7 in Table V.

TABLE V. LIST OF DELAYS BETWEEN EVENTS IN DATA SET 7

Delay(s)	Count	Delay(s)	Count
$d \leq 10$	27	$10 < d \leq 60$	32
$60 < d \leq 160$	10	$160 < d \leq 300$	11
$300 < d \leq 600$	3	$600 < d \leq 1200$	8
$1200 < d \leq 2400$	5	$2400 < d \leq 3600$	2
$3600 < d \leq 7200$	4	$7200 < d \leq 10800$	2
$10800 < d \leq 14400$	0	$14400 < d \leq 18000$	0
$18000 < d \leq 36000$	2	$36000 < d$	29

According to this table, we found 1115 pauses and therefore 1116 events. According to our 80/20 rule, we say that the lowest time delays are too short to determine a intermission between two operator sequences. So all time delays, which are longer or equal than160 s separate two operator sequences. Because in the other Datasets using this ratio exceeds the boundaries, we determined in the last Section, we also used a ratio of 50/50, which resulted in a pause of 10 s.

C. Elbow method

For this method, we start with a pause of 10 s. We subsequently increase this by another 10 s. Now the sequences are generated for every step and then they are counted. The count of sequences is plot, as shown in Figure 5. We now search for a horizontal area within the boundaries defined in Section III-A. We use either the first one, which stays ten times horizontal or the longest within the boundaries. The start of the resulting horizontal area is the resulting pause.

D. Comparison of the methods and Discussion

Now, we perform the methods described above on all our Datasets. The results are shown in Table VI. For all but one datasets the 80:20 ratio resulted in a bigger value than the upper bound, we determined. From all of the results, which are in between the boundaries, we choose the smallest number further on. We will use this further on for this work.

TABLE VI. COMPARISON BETWEEN THE METHODS TO DETERMINE THE LENGTH OF THE OPERATOR SEQUENCES

Method	Dataset		
	5	6	7
80:20	160	26322	58807
50:50	10	290	162
Ellbow	1160	950	420
Choosen value	160	290	162
Resulting Sequences	162	203	59

IV. Similarity of Fingerprints

After, we have determined how operator sequences can be created, we will improve our fingerprint data. In a first step, we grouped the fingerprints by their respective operator sequence. Now, we expected that the fingerprints inside a group are similar to each other, and to the fingerprints of other

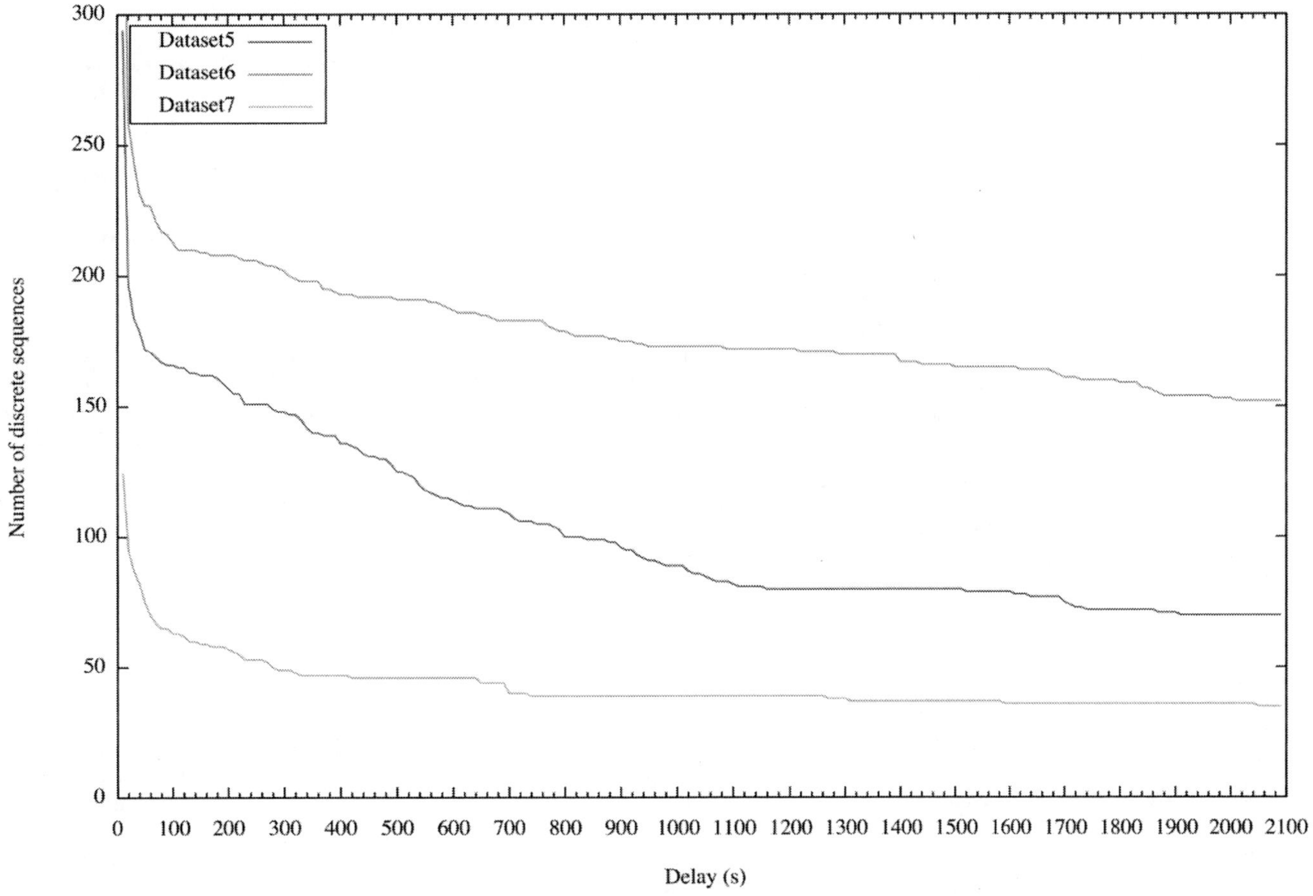

Figure 5. Results for the elbow method

groups they are dissimilar. So our target is to get our data more selective. This will be done by choosing the most useful distance measure for each dataset, then filtering the important columns and finally testing different fingerprint lengths. The results of these steps will be printed afterwards.

A. Preparing the data

Before, we start improving our data, we do some preparing steps. For example, we sort out some unused or cumulative columns. And we perform normalization because of the different ranges of the columns so every one is in the range between 0 and 1.

1) Determining unnecessary columns: As, we investigated our data further, we found three types of columns, which seem not to be helpful and therefore, should not be used further on:

- Columns, which are not used, e.g., having a null value
- Columns having only one value over the whole time
- Cumulating columns, e.g., we have a column counting the energy the machine needed

2) Normalization of columns: The next issue, we discovered is that ranges, e.g.:

- Temperate ranges from $\approx 0°C$ to $\approx 250°C$
- Percentage values (OP), from 0 to 100
- Pressure values from 0 to 60
- Button or Binary values from 0 to 1

Using this normalization, we get a range between 0 and 1 for all columns (Equation 1).

$$x'_i = \frac{x_i - mean(\vec{x})}{max(\vec{x}) - min(\vec{x})} \tag{1}$$

B. Optimization of the Fingerprints

One of our major goals is to achieve the improvement of the fingerprints autonomous. To achieve this, we have to define a quality-measurement. For every case, we have a pair of a key (fingerprint) and value (operator sequence). According to the values they can be divided in the following cases:

- both pairs have the same operator sequence: d_{int}
- both pairs have different operator sequences: d_{ext}

Now, we assume that a good fingerprint will have a small d_{int} and a big d_{ext}. Let i be the number of results for d_{int} and e the number of results for d_{ext}, we can now define two ways to express our quality measure:

$$d_{sum} = \left(\sum_{n=1}^{e} d_{ext}\right) - \left(\sum_{n=1}^{i} d_{int}\right) \tag{2}$$

TABLE VII. AVAILABLE DISTANCE MEASUREMENTS

Point set distances	**Point distances**
Hausdorff distance	Manhattan distance
Average Linkage	Chebychev distance
Single Linkage	Euclidean
Maximum Linkage	Squared Euclidean

and

$$d_{avg} = \frac{\left(\sum_{n=1}^{e} d_{ext}\right)}{e} - \frac{\left(\sum_{n=1}^{i} d_{int}\right)}{i} \qquad (3)$$

For now on, we will use the d_{avg} (Equation 3).

1) Choosing the most suitable Distance Measurement: Now, we want to determine the best working distance measurement. As described in our previous work [1], we use the distance measurements as listet in Table VII.

For every possible combination of point set and point distances, we determine the quality. As the different distance measures will result in different value ranges, we also have to normalize this results. The combination resulting in the highest quality will be used. The results for all three Datasets are shown in Table VIII.

2) Filtering the Columns: After, we have determined a suitable distance measurement, we are faced with the question if we only have relevant columns in our data. We assume, that using all columns will lead to a worse result than using only some.

At first, we reduce the fingerprint to one column. For this column d_{avg} is determined. This step is repeated for all other columns in the data.

The next step is a Greedy like algorithm. We sort the results from the last step. The column with the highest value is set for the resulting fingerprint. Now, we subsequently add the next columns and again determine d_{avg}. If the new quality is at least 5% better than the old one, the column will be added to the fingerprint.

After doing this for all columns, we finally get a set of columns for our resulting fingerprint. Beside the columns found in this algorithm the fingerprint also contains the columns containing data regarding operator events and columns, we need to determine wether the machine is manufacturing the product or not. As for the last step the results for this improvement step are also shown in Table VIII.

3) Fingerprint Length: In our last papers, we set the fingerprint length to exactly 5 minutes. We always struggled with this decision and wanted to verify, if this would be a suitable number. We now want to determine a suitable length for the fingerprint. At a first step, we should specify a upper and a lower bound. The lower bound is the smallest possible length of a fingerprint. As, we get our data every 10 seconds, the smallest possible fingerprint is 10 seconds long. The upper bound will be set to 30 minutes for now. The quality for every range is determined. The timespan with the best quality will be used. The results for this step are also shown in Table VIII.

C. Results

Table VIII shows the results for the improving steps. Now the matrix plots shown in Figures 6, 7, 8 are created for every dataset and every improvement step to show how our improvement has worked out. The color of each cell shows the similarity between two fingerprints. A green color shows a close distance, the red color a far distance. The operator sequences are grouped. So inside a white rectangle they all fingerprints have the same operator sequence. The perfect result would be green groups lying on the diagonale and red groups elsewhere. An Additional matrix plots, ordered by time is added.

According to the results in Table VIII for two Datasets, a short fingerprint was a better choice. The distance measurement is also the same for every Dataset.

1) Results on Dataset 5: Table IX lists all operator sequence which, we have found in Data set 5. The Shortcuts are used as labels in Figure 6. We have found overall 125 sequences with 21 different operator sequences. The step with the most difference seem to be the step using the Chebychev distance instead of the Euclidean distance. Also the step shorting the fingerprint makes a noticeable difference. Overall, we seem to get more red squares beside the diagonal line, this seems to be a step in the right direction. We are not satisfied with this result at all, because, we do not have green rectangles on the diagonale, as, we hoped

2) Results on Dataset 6: In Table X lists all operator sequence which, we have found in Data set 6. The Shortcuts are used as labels in Figure 7. We have found overall 145 sequences with 14 different operator sequences. As in Dataset 5 the step with the most difference seem to be the step using the Chebychev distance instead of the Euclidean distance. In this dataset, we noticed that a bigger fingerprint performs better, but, we do not see such a noticeable difference in the result as in Dataset 5. With this results, we are also not satisfied yet.

3) Results on Dataset 7: In Table XI lists all operator sequence which, we have found in Data set 7. The Shortcuts are used as labels in Figure 8. We have found overall 21 sequences with nine different operator sequences. Dataset 7 has the most interesting results. Unfortunately it is also the smallest data set, regarding to the number of operator sequences, we found. Changing the distance to Chebychev resulted in a green diagonale and now only the elements lying on this diagonale seem to be similar. The next step, filtering the columns anyway seems to improve the result a bit. The final step, changing the fingerprint length, also changed the result. We are not sure if, this is an improvement or not. Although this Dataset let us hope that, we can achieve some improvement of our data with the algorithms described in this work.

V. CONCLUSION AND FUTURE WORK

As described in Section IV, we tried to improve our results by optimizing the data, we gathered from the production machines. For nearly all of our improvement steps, we found a more or less noticeable difference in the results. The most distinct change was for changing the distance measurement. In this case the point set distance stayed at the Hausdorff distance, but the point distance was changed from the Euclidean to Chebychev distance. We have a lot of work to do, because our results are yet not satisfying. As we noticed some changes, especially regarding to Dataset 7, we think we are on the course to get some improvements in the future.

By now the sequence in which, we perform the improvement steps is arbitrarily chosen. By changing this order or

TABLE VIII. RESULTS OF THE OPTIMIZATION ALGORITHMS

Optimizations	Dataset 5	Dataset 6	Dataset 7
Columns original	120	225	177
Useless columns	47	75	56
Point Set Distance	Hausdorff	Hausdorff	Hausdorff
PointDistance	Chebychev	Chebychev	Chebychev
Columns	Tool.Temp2.OP Extruder.Temp0.OP Extruder.Temp4.OP Vacuumtank.In.Vacuum1.Val Meltpump.In.RPM.Band_Min Vacuumtank.In.Vacuum0.Val Tool.Temp0.OP	General.In.Power.Band_Min Extruder.In.Power.Band_Min General.In.Power.Val Vacuumtank.In.WaterTemperature3.Band_Max Vacuumtank.In.WaterTemperature1.Val Tool.Temp9.PV Tool.Temp19.PV Tool.Temp5.OP Tool.Temp18.SP	N2_PDO_Istwert1[1] N2_PDO_Istwert4[1] PDO5_PLCAnalog_In11_12[0] N1_PDO_Istwert3[4] N1_PDO_Istwert4[4]
Length (s)	70	1850	30

TABLE IX. OPERATOR SEQUENCES FOUND IN DATA SET 5

Shortcut	Sequence	Count
S1	Do nothing	2
S2	Meltpump.Out.Val_-0.1#	9
S3	Meltpump.Out.Val_-0.2#	8
S4	Meltpump.Out.Val_-0.3#	7
S5	Meltpump.Out.Val_-0.4#	4
S6	Meltpump.Out.Val_-0.5#	4
S7	Meltpump.Out.Val_-0.6#	2
S8	Meltpump.Out.Val_-0.7#	2
S9	Meltpump.Out.Val_-0.8#	2
S10	Meltpump.Out.Val_-1.0#	4
S11	Meltpump.Out.Val_0.1#	16
S12	Meltpump.Out.Val_0.2#	23
S13	Meltpump.Out.Val_0.3#	12
S14	Meltpump.Out.Val_0.4#	7
S15	Meltpump.Out.Val_0.5#	6
S16	Meltpump.Out.Val_0.6#	5
S17	Meltpump.Out.Val_1.0#	4
S18	Meltpump.Out.Val_1.7#	2
S19	Meltpump.Out.Val_2.0#	4
S20	Meltpump.Out.Val_6.9#	2
S21	Tool.Temp3.SP_10.0#	2

TABLE X. OPERATOR SEQUENCES FOUND IN DATA SET 6

Shortcut	Sequence	Count
S1	Do nothing	2
S2	Meltpump.Out.Val_-0.1#	20
S3	Meltpump.Out.Val_-0.2#	21
S4	Meltpump.Out.Val_-0.3#	4
S5	Meltpump.Out.Val_-0.4#	7
S6	Meltpump.Out.Val_-0.5#	4
S7	Meltpump.Out.Val_-0.6#	4
S8	Meltpump.Out.Val_-0.8#	2
S9	Meltpump.Out.Val_0.1#	17
S10	Meltpump.Out.Val_0.2#	28
S11	Meltpump.Out.Val_0.3#	21
S12	Meltpump.Out.Val_0.4#	12
S13	Meltpump.Out.Val_0.5#	2
S14	Meltpump.Out.Val_0.6#	3

TABLE XI. OPERATOR SEQUENCES FOUND IN DATA SET 7

Shortcut	Sequence Count	
S1	Do nothing	4
S2	N1_SP0_10.0#N1_SP1_10.0# N1_SP2_10.0#N1_SP3_10.0#	2
S3	N1_SP0_20.0#N1_SP1_20.0# N1_SP2_20.0#N1_SP3_20.0#	2
S4	Analog_Out_-2.0#	2
S5	Analog_Out_-8.0#	2
S6	Analog_Out_1.0#	2
S7	Analog_Out_10.0#	2
S8	Analog_Out_20.0#	3
S9	Analog_Out_7.0#	2

repeating some of the steps we might get a further improvement. E.g., if we determine the fingerprint length first, then perform column optimization and finally improve the fingerprint length again. Also, we should run the column filtering with all possible distance measurements combinations. This might result in a better set of columns then using only one.

In Section IV, we also mentioned that, we use 5% as a threshold to detect whether a improvement in quality is significant or not. We should question this value and evaluate if a smaller or bigger threshold can improve the results. Also, we currently use the operator event data for both, fingerprint and operator sequence. We should also verify if it is better to use them only if they are detected by our improvement algorithm.

To verify the result, we can, e.g., use the method, we developed in our previous work [1] together with the matrix plots, we should be able to determine whether, we get better results or not.

Another problem, we have with some improvement steps is the long computing time. So, we have to discover how much data, we have to have to get a proper improvement. Or how often the system should run the steps (e.g., once a week or once a month). After performing the improvement steps, the system has to create the knowledge base again, so, we should keep this in mind.

The retrieval methods we currently use to get recommendations out of our knowledge base, do not consider the order of the data points. In our future work, we will also consider other distance measurements, which take care about the order of the points. One of our current ideas is taking the data no longer as a n-dimensional point-set but consider it as a n-dimensional polygonal chain.

An advantage of the described algorithms is that they do not need any information about the process. So therefore they are not influenced by the operator and can easily be adopted to other process types. This is important because the transferr our system to other processes will be a major part of our future work.

REFERENCES

[1] V. Plenk, S. Lang, and F. Wogenstein, "An approach to provide user guidance in special purpose machines and its evaluation," International Journal On Advances in Software, vol. 10, no. 3 and 4, December 2017, pp. 167–179.

[2] V. Plenk, "Improving special purpose machine user-interfaces by machine-learning algorithms," Proceedings of CENTRIC 2016: The Ninth International Conference on Advanves in Human-oriented and Personalized Mechanisms, Technologies and Services, August 2016, pp. 24–28.

[3] V. Plenk, S. Lang, and F. Wogenstein, "Scoring of machine-learning algorithms for providing user guidance in special purpose machines," October 2017, pp. 7–13.

[4] N. Ben Rabah, R. Saddem, F. Ben Hmida, C.-M. V., and M. Tagina, "Intelligent case based decision support system for online diagnosis of automated production system," 13th European Workshop on Advanced Control and Diagnosis (ACD 2016), 2016.

[5] C. Ullrich, M. Aust, M. Dietrich, N. Herbig, C. Igel, N. Kreggenfeld, C. Prinz, F. Raber, S. Schwantzer, and F. Sulzmann, "Appsist status-bericht: Realisierung einer plattform für assistenz-und wissensdienste für die industrie 4.0." in DeLFI Workshops, 2016, pp. 174–180.

[6] N. Cen, K. Faisal, and I. M. Tariq, "Real-time fault diagnosis using knowledge-based expert system," Process Safety and Environmental Protection 86, no. 86, 2007, pp. 55–71.

[7] J. Lunze, Ereignisdiskrete Systeme. München: Oldenbourg, 2006.

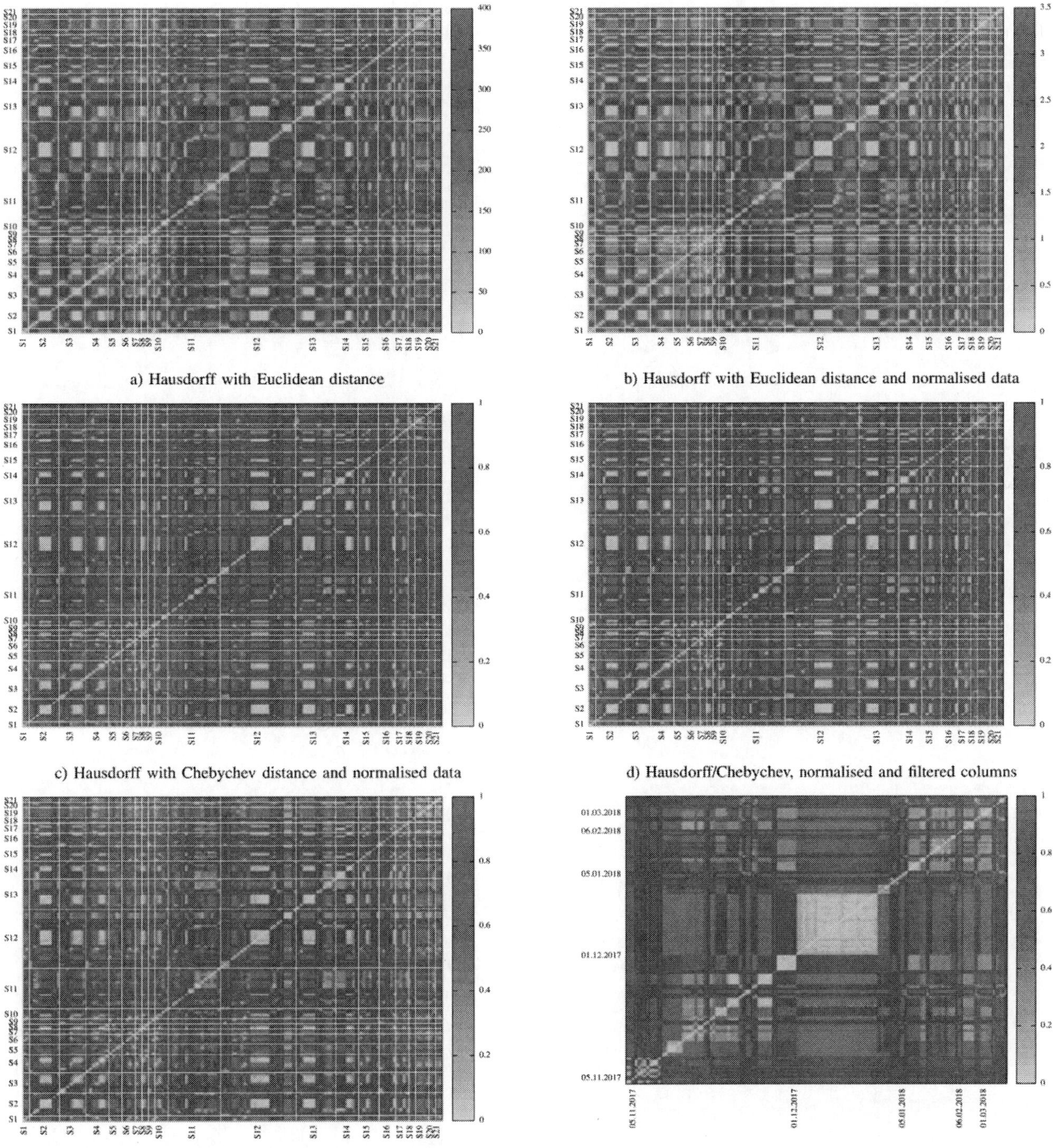

a) Hausdorff with Euclidean distance

b) Hausdorff with Euclidean distance and normalised data

c) Hausdorff with Chebychev distance and normalised data

d) Hausdorff/Chebychev, normalised and filtered columns

e) Hausdorff / Chebychev, normalised and filterd columns and longer fingerprint

f) Same settings as e) but ordered by time

Figure 6. Results for Dataset 5

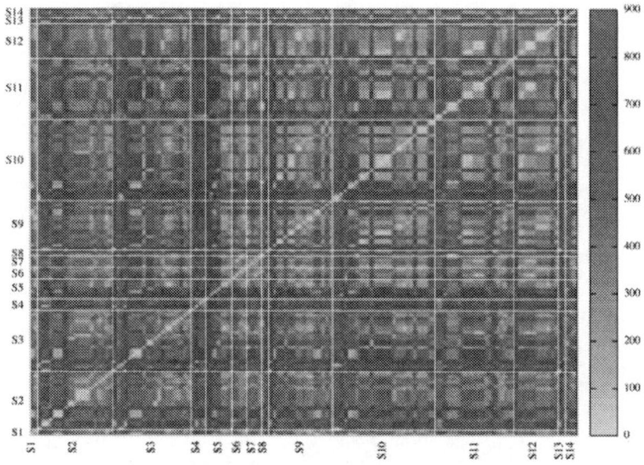

a) Hausdorff with Euclidean distance

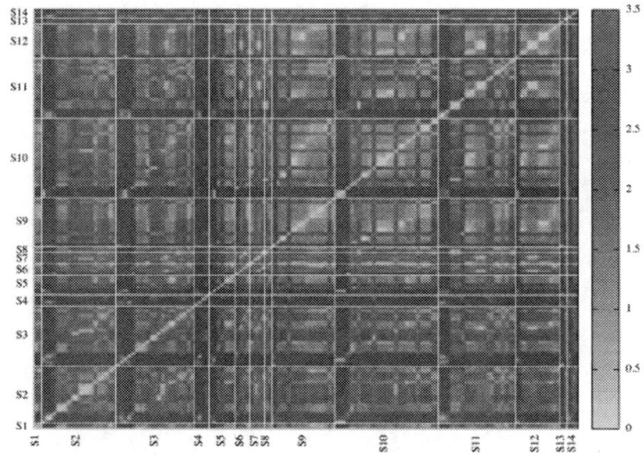

b) Hausdorff with Euclidean distance and normalised data

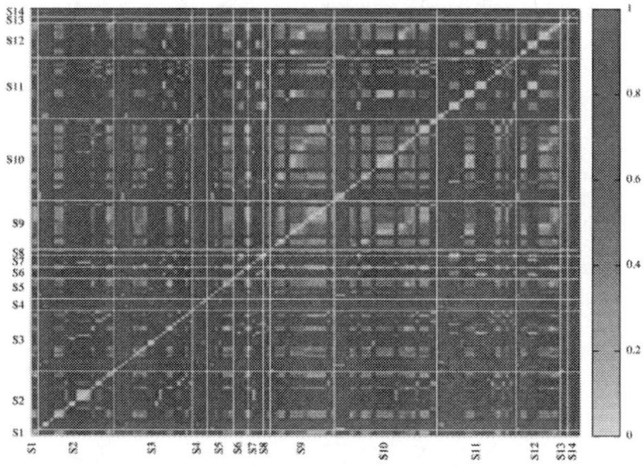

c) Hausdorff with Chebychev distance and normalised data

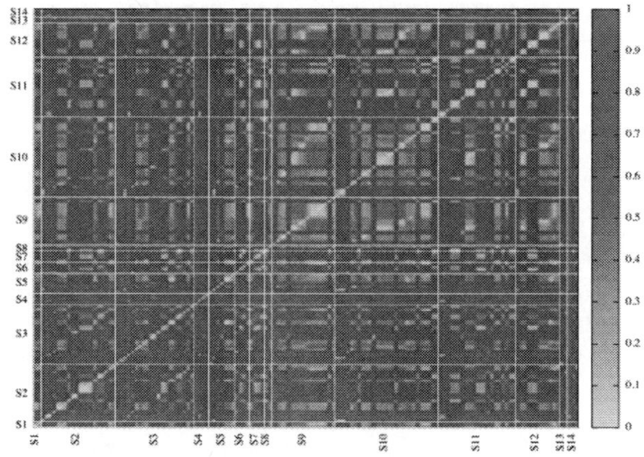

d) Hausdorff/Chebychev, normalised and filtered columns

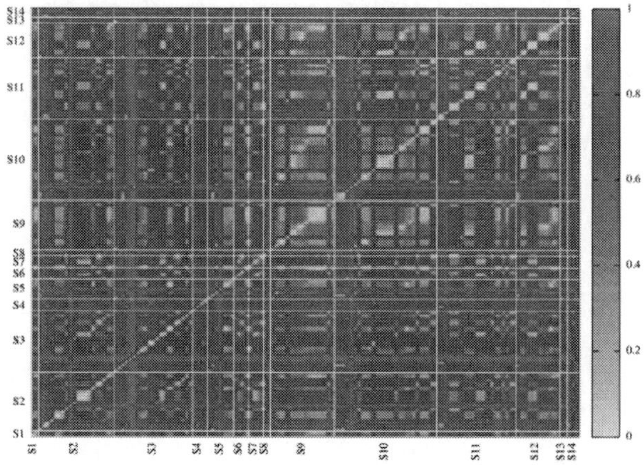

e) Hausdorff / Chebychev, normalised and filterd columns and longer fingerprint

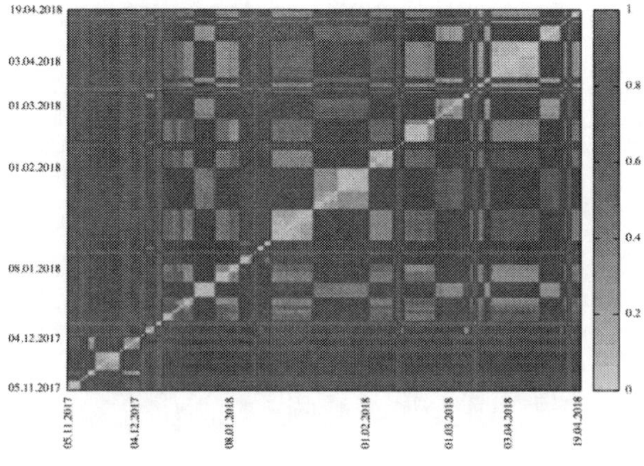

f) Same settings as e) but ordered by time

Figure 7. Results for Dataset 6

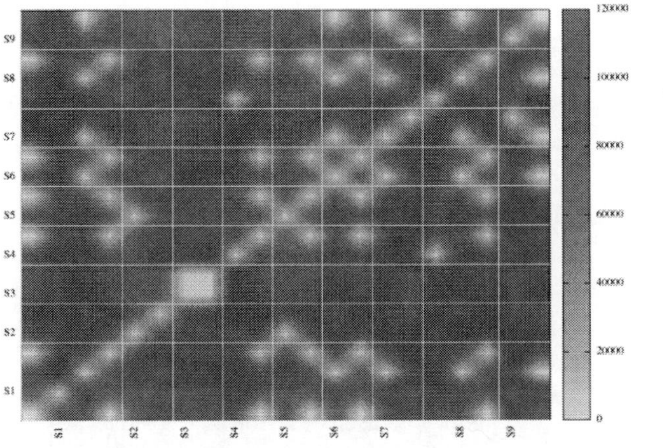

a) Hausdorff with Euclidean distance

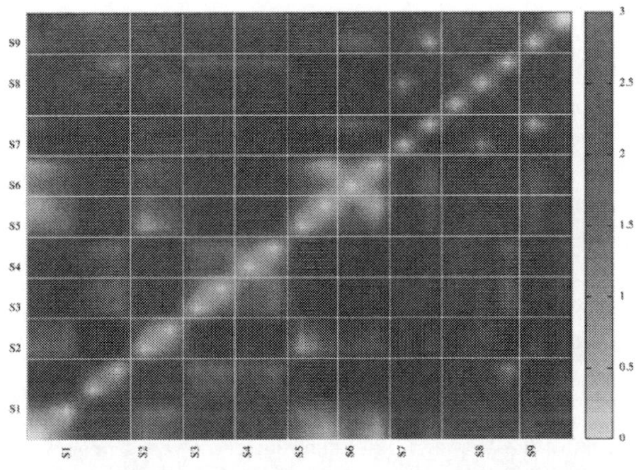

b) Hausdorff with Euclidean distance and normalised data

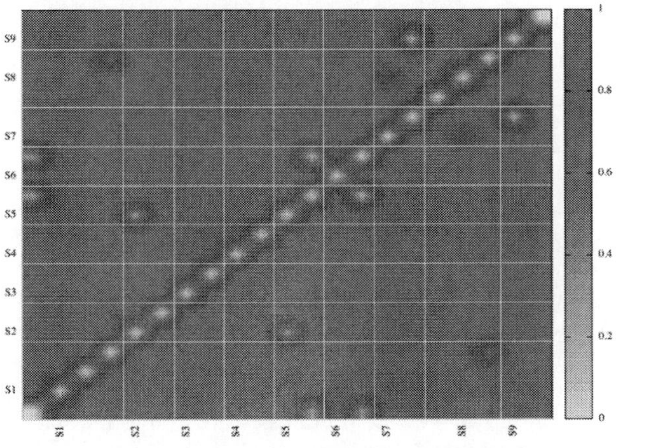

c) Hausdorff with Chebychev distance and normalised data

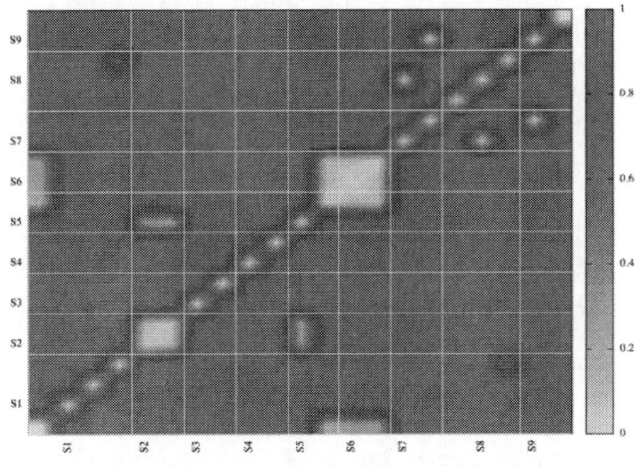

d) Hausdorff/Chebychev, normalised and filtered columns

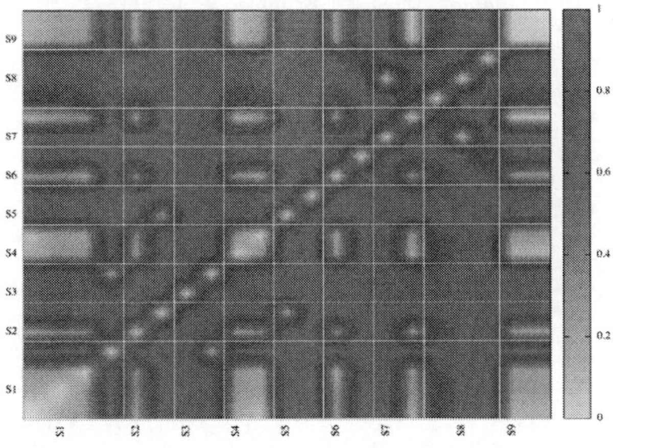

e) Hausdorff / Chebychev, normalised and filterd columns and longer fingerprint

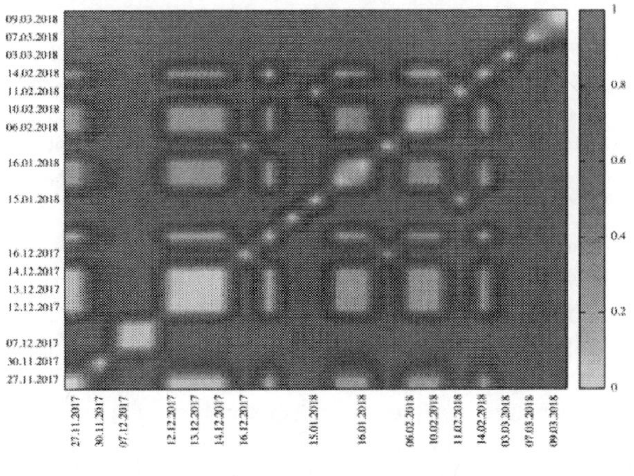

f) Same settings as e) but ordered by time

Figure 8. Results for Dataset 7

User-friendly Visualization of Energy Flows in Smart Homes

Thomas Dressel
Institute of Information Systems
Hof University of Applied Sciences
Hof, Germany
email: thomas.dressel@iisys.de

René Peinl
Institute of Information Systems
Hof University of Applied Sciences
Hof, Germany
email: rene.peinl@iisys.de

***Abstract*—The constantly increasing energy consumption requires to make people aware of energy and the trend towards the smart home makes it possible. Every manufacturer of so-called smart products uses its own application to display the data obtained. However, this makes it difficult for users to see a connection between individual devices. Even if people are using a smart home hub to collect data from multiple devices, the data is still separated in different charts or tables and hard to compare. To offer them more transparency in their smart home, an application was developed for electricians that want to offer their customers added value without great effort. This app makes the data of the devices available in a central location and provides an overview of the energy flow users can easily interpret.**

***Index Terms*—Smart home, energy monitor, energy flow.**

I. Introduction

Energy consumption has been rising for years and has more than doubled between 1990 and 2017 [1]. Despite significant improvements in energy efficiency, the electricity consumption in the European Union per household has increased by 2% per year from 2000 until 2010 [2]. Reasons for this are manifold and country-specific. They include more widespread use of electronic equipment like electric dryer and dishwasher that go along with a higher degree of basic comfort [2], more entertainment electronics like large tv sets [3], gaming PCs and video game consoles, as well as extensive use of air conditioning and cooling devices. Energy consumption of Information and Communication Technologies (ICT) and Consumer Electronics (CE) has risen considerably in recent years and now accounts for approximately 15% of global residential electricity consumption. Research further indicates that almost 41% of supplied power is wasted, e.g., through devices in standby or wrong room temperature [4].

The emergence of modern smart home devices plays an ambivalent role since these devices require additional power on the one hand, but also offer a certain benefit and potential for saving energy on the other hand [5]. These smart devices provide the user with a wide range of data. This includes both the user's energy consumption and energy generation through various devices. The available data of the individual networked devices are usually only displayed in a smartphone application of the respective manufacturer. As a result, you have to use several apps to get all the information. A disadvantage of this is that the data are separate from each other and a possible connection is difficult to recognize. In order to offer the user more transparency for its energy consumption or its energy production and the devices used, the data must be visible in a central location. Several studies show that this kind of feedback can change user behavior in a positive way [6]–[9]. This is especially important, since user behavior is a relevant factor besides the devices mentioned before [10] and people also have certain beliefs regarding energy consumptions and possible savings that are far away from the objective facts [11]. To overcome some of these problems, the Energy Flow Monitor (*EFM*) was developed, a smartphone application that visualizes data from different devices of different manufacturers centrally in one app. Modern households include a variety of energy consuming and also generating devices like heat pumps, solar heating, photovoltaic collectors. Therefore, it is increasingly hard for users to keep track of their interplay. Is the dryer running while PhotoVoltaic (PV) collectors are providing electricity, or is it using expensive power from the national grid? *EFM* can show interactions like this to the user in an understandable manner. As a result, the user becomes more aware of how some devices are connected and where there is potential for energy saving.

The remaining paper is organized as follows. First, related work is presented and deficiencies of existing solutions are highlighted. This leads to the goal of our research, which is presented in section three. A short excursion to technical aspects of the developed app is discussed in section four, before our solution to the challenge is presented in section five. We conclude the paper with a discussion of limitations an outlook to future work and a conclusion.

II. Related Work

The consumption of energy is abstract, invisible, and untouchable - unlike most other consumable goods [12]. Therefore, home energy usage often goes unnoticed. The only means of judging their household energy usage is often a monthly (or bi-monthly) bill (ibid.). In Germany, it is even worse, since many people only get yearly bills.
Although there is both great potential and great interest in stimulating behavioral change with energy monitoring tools, we still know little about how to design, situate, and integrate them to help people make good decisions to safe energy

[13]. "Current applications often require too much effort and management from occupants and are not integrated well with the set of information tools people already use" (ibid). Bartram et al. propose a combination of Web application to display on the PC, mobile app with a subset of the features of the Web app (see Figure 1) and an ambient display in the kitchen to positively influence the occupants.

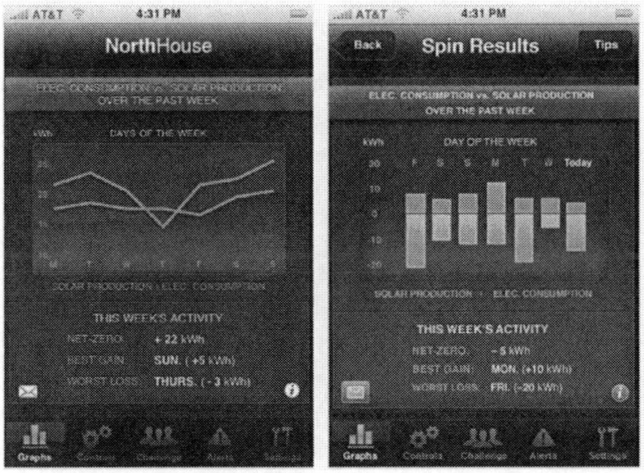

Fig. 1. Screenshot of the mobile app from Bartram et al. [13].

Existing studies paint a diverging picture of the impact of energy monitoring solutions on energy savings. [14] found no differences in energy consumption for one device and only insignificant differences (12%) for the other device. However, this could be due to their limited study size since 12% would be significant for a larger sample based on the measures used in the paper. The amount is also in line with findings in [15] that report 10% savings. [16] found that initially, energy monitors led to nearly 8% savings, but after four month, no difference could be observed anymore. However, in large meta-studies it was shown that feedback systems are effective in supporting energy saving behavior and resulted in about 10% savings [12], although individual study results differed between 5% and 15%.

One problem of most existing solutions is that they do either display only the overall consumption or integrate the consumption of individual devices using plug adapters, as e.g., [17] do. This excludes devices without a plug like oven, water heaters, PV collectors and heat pumps. Some of the latter devices come with monitoring apps on their own, but as stated before, an integrative view is missing. [18] claim that they can monitor individual devices like refrigerator, washing machine, dishwasher and dryer in addition to furnace and water heater based on the home's circuit breaker box (see Figure 2). However, this seems not plausible since usually there are only few fuses per house and only high power devices like the water heater and oven have a dedicated fuse, whereas refrigerator, washing machine and dishwasher share a fuse with other devices in the kitchen or cellar.

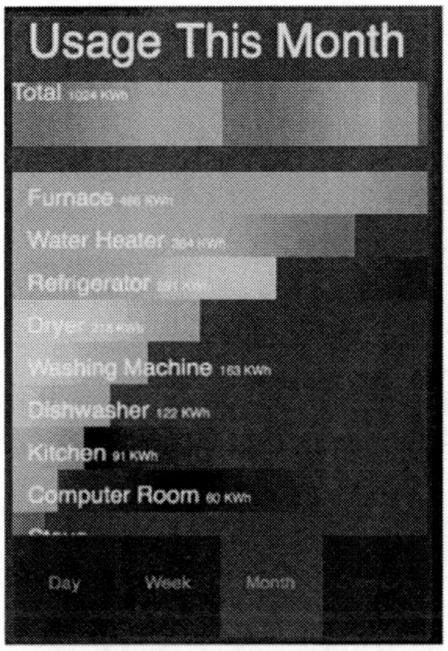

Fig. 2. Concept rendering of the *WattBot* application [18].

Feedback on energy consumption has a significant role in raising awareness. It can sensitize users to the extent that a reduction in consumption in the order of 10% is possible [6], [7]. One approach to communicate energy consumption to the user is by Fleisch et al. [19]. This system allows the user to monitor its energy consumption on a mobile phone and to compare devices with each other. The disadvantage of this solution is that one has to run its own web server, which may discourage some users from using it. There are also commercial solutions [20], which are installed on the electricity meter and offer the user a good overview of the electricity consumption, but there is no visualization how different devices are connected or the possibility to see the energy generation.

III. GOAL

In order to create more transparency for the user, an easy-to-understand overview of the main devices is important. These must not be scattered over several apps but integrated in one place. The resulting overview should provide the user with all important information and especially show the connection between the individual devices. The aim is to develop a mobile app that displays data of smart devices centrally and also provides an overview of the energy flow in a user-friendly way. This creates greater transparency and makes it easier for users to understand the energy consumption in their homes. User-friendly is used here in the sense that Trenner subsumes under "how user-friendly is commonly understood" [21]. That means "easy to use", "designed with the needs of users in mind" and "to serve the non-expert or novice" (ibid.). Expert users are not considered in the first iteration.

The application should also be adaptable for companies without great effort by choosing systems that are as open as possible. Since there are two major smartphone platforms in

Germany that both have more than 25% market share and together provide for over 95% of the market [22], the mobile app should be available for both platforms.

IV. Cross Platform App Development

Given the task at hand, the decision is between developing two native apps or one cross platform app. Native apps offer some advantages. They have full access to all features of the device and the operating system. In addition, they usually provide better performance and smoother animations [23]. A native user interface uses the original widgets provided by the operating system, and will also be updated over time with operating system updates. On the other hand, one has to develop two separate applications, using different programming languages, development environments and paradigms. Offering the application natively on two platforms will multiply the effort by factor 1.3 [24] or even up to factor two [23].

Whether cross-plattform apps are a good alternative depends on the complexity and type of application [23]. A huge challenge is providing a consistent app behavior and still respecting the UI guidelines of the respective platform (ibid.), which is also important for the user experience, since users expect a certain degree of consistency across apps on their device. Cross platform applications can be developed with different approaches: Web, Hybrid, Interpreted, Cross-compiled and Model-driven approach [25]. Web apps run in the browser and are not installed using the platform's app store. Hybrid apps are developed with HTML, CSS and JavaScript like Web apps, but are packaged into platform-specific apps and run inside a Web view instead of the browser. They can use more native features through abstraction layers like *Phonegap/Cordova*. With interpreters like *Appcelerator Titanium*, developers can use a single language, which is than interpreted in a platform-specific container. It allows for native user interface components, but comes with performance degradation. Cross-compilers generate true native code out of a common codebase. *Xamarin* is probably the most popular example of this category. Model-driven development is a big trend, but it is not yet established in companies outside academia.

Having only web developers available that had no experience with *Android* nor *iOS* development, and the application being not too complex, a closer look at frameworks supporting the hybrid approach seemed sensible. This choice is also becoming more common since both performance and access to native features of hybrid apps has become better over the last years. [26] identified *Ionic* (hybrid) and *React Native* (cross-compiler) as innovative approaches and concluded that *Ionic* made development of interfaces and app flow easy through their component library, whereas *React Native* left more architectural choices to the developer. On the other hand, *React Native* mostly delivered un-styled and "not native-looking" interface components with the option of styling them to fit the app. *Ionic* provided a massive library of ready-made and pre-styled components making the development of standard native-looking interfaces easy (ibid.).

Recently, a new alternative became available, which is called progressive web apps [27]. Although they performed well in the comparison, we did not further consider them since they do not produce apps for *Android* and *iOS* app stores. Despite the fact that our own app does not make excessive use of native components, our choice was still *Ionic*, since other papers describe it as the "top choice among hybrid mobile app frameworks" [28]. The app presented here was therefore developed as a hybrid application using *Ionic*, *Angular* and *Apache Cordova* to increase its availability and to keep the programming effort as low as possible.

V. Own Solution

Froehlich found ten design dimensions when planning energy monitoring systems [12]: frequency, measurement unit, data granularity, push/pull, presentation medium, location, visual design, recommending action, comparison and social sharing. Our app displays current and historic energy consumption and flows with mostly a few seconds delay to recording. It uses kW as the measurement unit and displays data for PV collectors, heat pump, furnace and overall electricity consumption for a user-defined time interval (current, day, week, month). Additionally, it display the flow between national grid, heat pump, PV collectors and the electric devices in the house. Users have to actively query the data. There are no proactive push notifications currently. Information is displayed on mobile phones and therefore independent from the location of the monitored devices and even from anywhere on the world, as long as internet access is available. Visualization is not optimized for tablets given the dominance of smartphones and the comparatively small number of tablets in the market. We put much effort into the visual design, which consists of polished charts (see Figure 4) and our special energy flow visualization (see Figure 5). *EFM* does currently not recommend any actions, nor does it directly contain a social sharing functionality. Comparison with past data is easy, using the charts. Comparison with other households is not implemented.

Apache Cordova and *Ionic* were used for this application. Devices whose data can be displayed are a heat pump, PV collectors, heating, solar heating, thermal buffer, the total power consumption and energy accumulators. The data include the current consumption at the time of logging, the power gained or the temperature profile.

One important consideration for our solution was the ability to use it independently of the manufacturer of the heat pump or furnace. This gives electricians much more freedom in chosing the best solution for the customer, without thinking about the data connection. The latter is provided by sensors from a company called *Technische Alternative* [29]. They can be used in conjunction with all leading manufacturers. Despite the availability of a number of smart home hubs like OpenHAB (based on Eclipse Smart Home [30]), IP Symcon or others, including central heating systems, PV and water heaters is still a challenge for end users. Whereas lighting, blinds, temperature and other sensor data is usually available via standard protocols like Zigbee, Z-Wave or KNX, the listed devices usually do not provide an easy to use data interface. The Sensors by Technische Alternative can reduce this problem by providing energy data independent of the manufacturer and at affordable prices. A drawback is, that only certified electricians are allowed to install those sensors. This is much more effort compared to the usual plug-adapters that are merely plugged into the wall socket and the plug of the appliance is plugged into the adapter.

Avoiding the necessity to operate an own backend, a Mobile Backend as a Service (MBaaS) solution was desired [31] for retrieving the data via an Application Programming Interface (API). Fortunately, the *Technische Alternative* provides a web portal for accessing the data of its sensors in the field. In addition to the web interface for humans that is available for several years, they are offering an API for developers since spring 2018. We were granted early access to this API in order to develop EFM. Although this is not an MBaaS solution in a narrow sense, it provided all the necessary data. The company that provides the customer with this application therefore does not have to operate its own web server and thus has the advantage of offering the customer added value. In order to speed up visualization of data, it is stored in a local database on the phone after it was first retrieved from the backend. The database used for that is *SQLite*.

The great advantage of our application is that both the views of the individual devices and a combined view to illustrate the interaction of the devices are located in a central place. Hybrid cars were used as a basic idea and inspiration because they also display an energy flow, e.g., the interaction of the engine, its consumption and recuperation is displayed in a common view.

The app is divided into four views for energy and energy flow and the view of settings. Immediately after starting the app, the user sees the overview page with all connected devices (see Figure 3). The current value and energy of the current day is displayed for power generating or power consuming devices. For devices such as the heater, other data are displayed accordingly, such as whether the status of the heater and the current temperature. Each device is shown on an own tile on the main screen (see Figure 3). For users, that have less devices, only relevant tiles are displayed. The acronym EVI that can be found in Figure 3, is the abbreviation for the heat pump or our project partner.

If a device is selected in the overview by touching a tile, the detailed view for the respective device is shown (see Figure 4). In this detail view, you can see the energy production for the current course of the day, the current week, the current month and the current year.

In the overview of all devices, users will also find the tile “Energy Monitor”. If this is selected, a view is displayed showing the energy flow and thus the interaction in the house

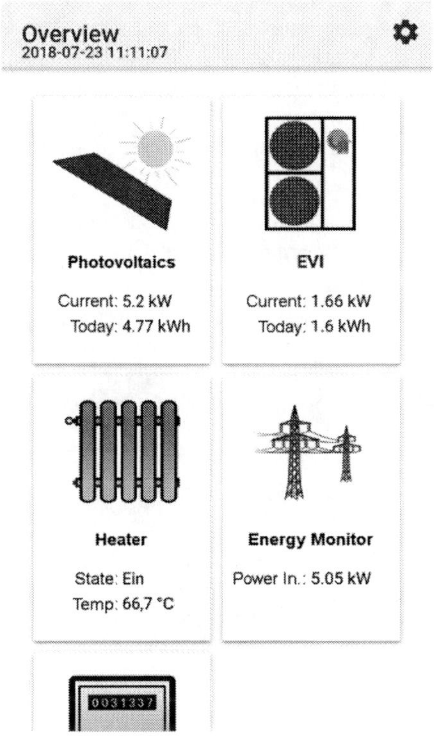

Fig. 3. Example overview showing the connected devices with current values.

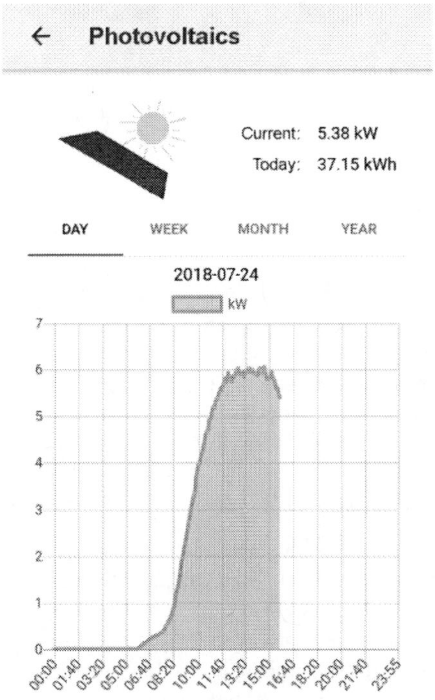

Fig. 4. The course of the current day can be seen in the detailed view.

(see Figure 5). If the user has a photovoltaic system, the current energy generation is displayed there. The total energy consumption and how it is divided into heat pump and other consumers can also be seen. If more energy is generated by the photovoltaic system than is consumed, it is fed into the national power grid and can also be seen in the energy monitor.

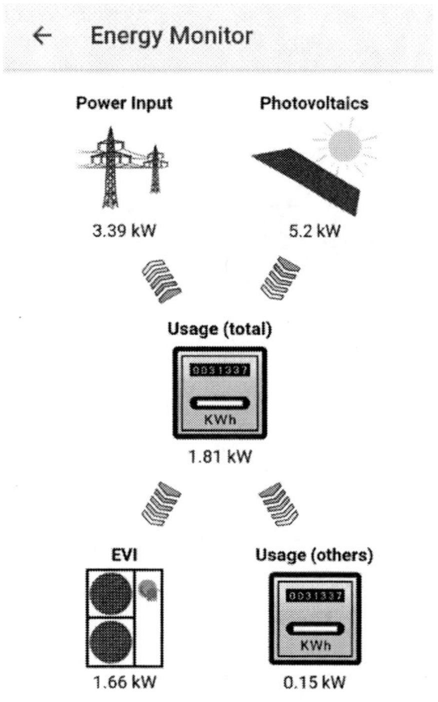

Fig. 5. The "Energie Monitor" shows the energy flow and the interrelationships of individual devices.

If the user has not yet set up the app, the settings are displayed directly at the first start (see Figure 6). The only information users are required to enter are username and password, which they know from their login to the Web portal of "Technische Alternative" as well as the CMI ID and the name of the profile that can be easily found in the Web portal. The App comes with preconfigured mappings from the most common devices used by the project partner to the displayed variables so that there is no configuration effort here.

VI. Limitations & Future Work

The current app does not incorporate all of the aspects Froehlich suggests [12]. The envisioned target group is not known to be excessive users of social media. However, enabling the sharing of energy consumption with friends seems an interesting improvement. It would be even more valuable, but also much more complex to include recommendation for actions that are specific to the sensor readings of the household monitored. Those recommendations must be realizable and not trivial. Otherwise, users will reject them and get a bad

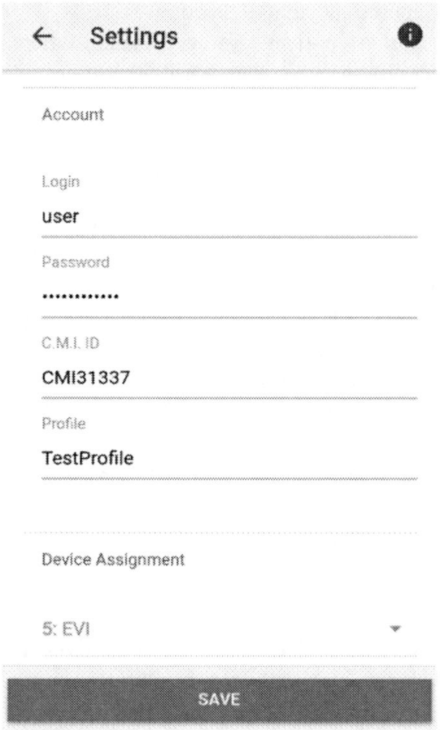

Fig. 6. The user has to enter its credentials, the corresponding identity of the device of the *Technische Alternative* [29], a created profile and the device assignment.

impression of the app. Due to the limited amount of time to develop EFM, this aspect was not implemented yet. Once the app has a considerable amount of users, it would be also interesting to offer comparisons between different households. These comparisons only make sense for households that are sufficiently similar. Comparing a three person household with 120 square meters living space to another one with five inhabitants and 180 square meters is nonsense. Therefore, the app needs to collect meta data about the household in order to support such a feature. [32] found that users' willingness to reduce their energy consumption can be increased from 25% to nearly 60% when appropriate comparison with peer groups is provided. However, the strict data protection rules of the European Union have to be taken into consideration, which might preclude the feedback with name and photo of other users that was used in the mentioned study. Finally, it should also be investigated whether the performance of the app can be further improved. We did not formal evaluation of the user-friendliness of the app, but initial feedback of a few test users was positive.

VII. Conclusion

This paper introduced *EFM*, a smartphone application that provides a central place for users to view data from their smart devices, especially power consumers and generators. It was developed as a hybrid app using *Ionic*, *Angular* and *Apache Cordova* and is able to run on *Android* and *iOS* devices. Thus, it is no longer necessary to use several different device- or manufacturer-specific apps to display device data. It collects data from different devices like heat pumps, PV collectors and furnace using sensors from *Technische Alternative*, independently of the manufacturer of those devices. Data is made available by *Technische Alternative* in both a web portal in human readable form and an API for consumption in own apps. Our app excels the web portal in providing an integrated view on energy flows between different devices, which is a significant benefit compared to single data of devices.

ACKNOWLEDGEMENT

This work has partly been funded by the European Regional Development Fund (ERDF) as part of the GreenTec Workshop Hof project. The authors would like to thank the anonymous reviewers, our industry partner Gemeinhardt AG, Manuela Wimmer and Valentin Plenk for their help.

REFERENCES

[1] Enerdata, "Global Energy Statistical Yearbook 2018." [Online]. Available: https://yearbook.enerdata.net/electricity/electricity-domestic-consumption-data.html retrieved: September, 2018.

[2] A. De Almeida, P. Fonseca, B. Schlomann, and N. Feilberg, "Characterization of the household electricity consumption in the EU, potential energy savings and specific policy recommendations," *Energy and Buildings*, vol. 43, no. 8, pp. 1884–1894, 2011.

[3] T. Crosbie, "Household energy consumption and consumer electronics: The case of television," *Energy Policy*, vol. 36, no. 6, pp. 2191–2199, 2008.

[4] R. J. Meyers, E. D. Williams, and H. S. Matthews, "Scoping the potential of monitoring and control technologies to reduce energy use in homes," *Energy and Buildings*, vol. 42, no. 5, pp. 563–569, 2010.

[5] A.-G. Paetz, E. Dütschke, and W. Fichtner, "Smart homes as a means to sustainable energy consumption: A study of consumer perceptions," *Journal of consumer policy*, vol. 35, no. 1, pp. 23–41, 2012.

[6] S. Darby, "Making it obvious: designing feedback into energy consumption," in *Energy efficiency in household appliances and lighting*. Springer, 2001, pp. 685–696.

[7] ——, "The effectiveness of feedback on energy consumption," Environmental Change Institute, University of Oxford, Oxford, UK, A review for DEFRA of the literature on metering, billing and direct displays, 2006.

[8] T. Ueno, F. Sano, O. Saeki, and K. Tsuji, "Effectiveness of an energy-consumption information system on energy savings in residential houses based on monitored data," *Applied Energy*, vol. 83, no. 2, pp. 166–183, 2006.

[9] W. Abrahamse, L. Steg, C. Vlek, and T. Rothengatter, "The effect of tailored information, goal setting, and tailored feedback on household energy use, energy-related behaviors, and behavioral antecedents," *Journal of environmental psychology*, vol. 27, no. 4, pp. 265–276, 2007.

[10] K. Steemers and G. Y. Yun, "Household energy consumption: a study of the role of occupants," *Building Research & Information*, vol. 37, no. 5, pp. 625–637, 2009.

[11] S. Z. Attari, M. L. DeKay, C. I. Davidson, and W. B. De Bruin, "Public perceptions of energy consumption and savings," *Proceedings of the National Academy of sciences*, vol. 107, no. 37, pp. 16 054–16 059, 2010.

[12] J. Froehlich, "Promoting energy efficient behaviors in the home through feedback: The role of human-computer interaction," in *Proceedings of the HCIC Workshop*, vol. 9, 2009, pp. 1–11.

[13] L. Bartram, J. Rodgers, and K. Muise, "Chasing the negawatt: visualization for sustainable living," *IEEE Computer Graphics and Applications*, vol. 30, no. 3, pp. 8–14, 2010.

[14] M. A. Alahmad, P. G. Wheeler, A. Schwer, J. Eiden, and A. Brumbaugh, "A comparative study of three feedback devices for residential real-time energy monitoring," *IEEE Transactions on Industrial Electronics*, vol. 59, no. 4, pp. 2002–2013, 2012.

[15] T. Ueno, R. Inada, O. Saeki, and K. Tsuji, "Effectiveness of displaying energy consumption data in residential houses. analysis on how the residents respond," in *Proceedings of the European Council for an Energy-efficient Economy*, vol. 6, 2005, p. 19.

[16] S. S. Van Dam, C. A. Bakker, and J. D. M. Van Hal, "Home energy monitors: impact over the medium-term," *Building Research & Information*, vol. 38, no. 5, pp. 458–469, 2010.

[17] A. Hashizume, T. Mizuno, and H. Mineno, "Energy monitoring system using sensor networks in residential houses," in *Proceedings of the 26th International Conference on Advanced Information Networking and Applications Workshops*. IEEE, 2012, pp. 595–600.

[18] D. Petersen, J. Steele, and J. Wilkerson, "WattBot: a residential electricity monitoring and feedback system," in *CHI'09 Extended Abstracts on Human Factors in Computing Systems*. ACM, 2009, pp. 2847–2852.

[19] M. Weiss, F. Mattern, T. Graml, T. Staake, and E. Fleisch, "Handy feedback: Connecting smart meters with mobile phones," in *Proceedings of the 8th International Conference on Mobile and Ubiquitous Multimedia*. ACM, 2009, p. 15.

[20] "Blue Line Innovations Inc." [Online]. Available: http://www.bluelineinnovations.com retrieved: September, 2018.

[21] L. Trenner, "How to win friends and influence people: definitions of user-friendliness in interactive computer systems," *Journal of information science*, vol. 13, no. 2, pp. 99–107, 1987.

[22] Statista, "Mobile Operating Systems - Market Share Internet Usage in Germany until 2018." [Online]. Available: https://de.statista.com/statistik/daten/studie/184332/umfrage/marktanteil-der-mobilen-betriebssysteme-in-deutschland-seit-2009/ retrieved: September, 2018.

[23] M. E. Joorabchi, A. Mesbah, and P. Kruchten, "Real challenges in mobile app development," in *Empirical Software Engineering and Measurement, 2013 ACM/IEEE International Symposium on*. IEEE, 2013, pp. 15–24.

[24] F. Ramisch, "Study: App development costs 30,000 euro on average," 2015. [Online]. Available: https://mobilbranche.de/2015/07/ibusiness-studie-die retrieved: September, 2018.

[25] M. Latif, Y. Lakhrissi, E. H. Nfaoui, and N. Es-Sbai, "Cross platform approach for mobile application development: A survey," in *Proceedings of the International Conference on Information Technology for Organizations Development*. IEEE, 2016, pp. 1–5.

[26] T. Majchrzak and T.-M. Grønli, "Comprehensive analysis of innovative cross-platform app development frameworks," in *Proceedings of the 50th Hawaii International Conference on System Sciences*, 2017.

[27] A. Biørn-Hansen, T. A. Majchrzak, and T.-M. Grønli, "Progressive web apps: The possible web-native unifier for mobile development," in *Proceedings of the 13th International Conference on Web Information Systems and Technologies (WEBIST)*, 2017, pp. 344–351.

[28] M. Q. Huynh, P. Ghimire, and D. Truong, "Hybrid app approach: could it mark the end of native app domination?" *Issues in Informing Science and Information Technology*, vol. 14, pp. 049–065, 2017.

[29] "Technische Alternative." [Online]. Available: https://www.ta.co.at/ retrieved: September, 2018.

[30] L. Smirek, G. Zimmermann, and M. Beigl, "Just a smart home or your smart home–a framework for personalized user interfaces based on eclipse smart home and universal remote console," *Procedia Computer Science*, vol. 98, pp. 107–116, 2016.

[31] K. Lane, "Overview of the backend as a service (baas) space," *API Evangelist*, 2015.

[32] M. Graffeo, I. Ritov, N. Bonini, and C. Hadjichristidis, "To make people save energy tell them what others do but also who they are: a preliminary study," *Frontiers in psychology*, vol. 6, p. 1287, 2015.

Smart Toys for Game-based and Toy-based Learning

A Study of Toy Marketers', Preschool teachers' and Parents' Perspectives on Play

Pirita Ihamäki
Prizztech Ltd.
Siltapuistokatu 14, 28101
Pori, Finland
pirita.ihamaki@prizz.fi

Katriina Heljakka
University of Turku
Po.box. 124, 28101
Pori, Finland
katriina.heljakka@utu.fi

***Abstract*—Over the next years smart Internet-connected toys are expected to grow significantly in numbers. Our study explores smart toys' potential to deliver experiences related to playful learning. One key aspect of toys, such as the CogniToys Dino, Fisher-Price's Smart Toy Bear and Wonder Workshop's Dash Robot are their game-based and toy-based features and functions, which are suggested to have educational outcomes when used in play. Through a comparative investigation of toy marketers', preschool teachers' and the parents' of preschool-aged children's perspectives of smart toys potential—and a comparison to the actual play experiences of preschoolers discovered in earlier stages of research, we demonstrate how the educational potential of contemporary smart toys may be categorized into game-based and toy-based affordances that may be employed for specific educational goals in playful learning.**

Keywords - game-based learning; Internet of Toys (IoToys); play; preschoolers; smart toys.

I. INTRODUCTION

This paper examines the kind of possibilities that the Internet of Toys (or, IoToys) offers for playful learning experiences through their various affordances analyzed through the Playful Experience (PLEX) model introduced by Lucero et al. [1]. We build on the suggested framework to conceptualize the gameful and toy play-like affordances of current smart toys. Our Playful Experience Framework for IoToys validation efforts include an earlier study of play-testing smart toys of the present to see what kind of play experiences these toys prompted in their players (in our case study, preschool-aged children). As a result, the 15 categories of the original framework were further divided into game-based and toy-based play experiences in the Playful Experience for IoToys Framework (See Table 1). This framework functions as a basis for our evaluation of the IoToys suitability for playful learning from the viewpoints of toy marketers, preschool teachers and preschooler's parents.

In earlier stages of our research, we have found that preschoolers use the toys for many kinds of play [2] [3]. This paper focuses on how these play experiences may be categorized into game-based and toy-based experiences, which both contribute to learning. With the previous findings in mind, our goal is to compare the results of the play tests with preschoolers with perspectives on smart toys as communicated by toy marketers, preschool teachers and parents of the preschool children who took part in our play tests. In this way, the paper at hand outlines three perspectives on playful learning related to smart toys: 1) the IoToys marketers' point of view, 2) preschool teachers' point of view, and 3) parents' point of view on the educational possibilities of these contemporary toys.

While some researchers believe that using computers before age seven subtracts from important developmental tasks and other types of learning [4], we believe that the IoToys with their 'hidden' [5], smart and connected technology, represents a medium for children's social and intellectual development. There have been effective attempts at educational technology integration, but school reformers often expect educators to know how to incorporate technology into their teaching. Whereas it may be difficult for educators to realize the potential of new kinds of smart toys to be used in the context of early education, it is important to increase knowledge of the toys' possibilities in the learning context. What is not to be neglected are the parents' attitudes and opinions of using smart toys, as they represent so-called 'edutainment', which may also be consumed when played with in the informal learning environment of the home. When children have a possibility to learn with smart toys, they can, for example, train their language and interaction skills. In this way, both the formal and informal playexperiences related to these toys change children's dispositions to learning and behaving.

According to an early definition, a 'smart toy' has been defined *as a device consisting of a physical toy component that connects to one or more toy computing services to facilitate gameplay in the cloud through networking and sensory technologies to enhance the functionality of a traditional toy* [6]. There are three general properties of a smart toys: They are 1) *pervasive* – a smart toy may follow the child through everyday activities, 2) *social* – social aspects and multiplayer participation are becoming a mandatory aspect of interactive IoToys in one-to-one, one-to-many and many- to-many relations [7], and 3) *connected* – contemporary smart toys as IoToys may connect and communicate with other toys and services through networks. For example, the CogniToys Dino can listen and answer to questions raised by children by voice recognition as the toy is connected to IBM Watson's Elemental Path "friendgine", which is a child friendly database [8]. Recently, the smart, but previously non-connected toys have evolved into connected playthings, that according to Holloway and Green (2016) mean IoToys, which 1) are connected to online platforms through Wi-Fi and Bluetooth, but can also connect to other toys, 2) are equipped with sensors, and 3) relate one-on-one to children [9].

According to the marketers' of IoToys, toys employed in our study like the CogniToys Dino, Wonder Workshop's Dash and Fisher-Price's Smart Toy Bear offer opportunities for both entertainment and learning. The toys present educational possibilities, particularly in informal situations. For example, the toy markers' promote the toys' capabilities to teach language when played with at home. In the toy marketers' understanding, then, the IoToys offer significant educational benefits and bring with them new possibilities to existing learning technologies [2].

Our interest to explore the potential of the smart toys in a formal learning environment led us to investigate the toys employment in the preschool context. After studying the play experiences of young children between the ages 5-6 years, we wanted to explore how these experiences correlate with the toy marketers', preschool teachers' and parents' perspectives of smart toys' affordances suited for learning. What guided our interest was the understanding that play with smart toys may be viewed from the perspectives of rule-bound (game-like) and open-ended (toy play-like) interaction. In other words, they provide opportunities for both game-based and toy-based learning.

In order to study the perspectives that the different stakeholders around smart toys have, we used three kinds of research material: 1) marketing texts related to the toys under investigation as communicated by IoToys marketers, and 2) a thematic survey to explore preschool teachers' and, 3) parental attitudes, and parents' experiences of digitally enhanced, smart toys in general. We interviewed two preschool teachers and the 14 parents of the altogether 20 preschoolers who participated in our playtests with the help of semi-structured surveys. The participants were asked about a wide range of topics in relation to digitally-enhanced toys, which we here consider to represent smart toys.

In the following section 2, we describe earlier studies to explain how marketers', early educators' and parents' attitudes have been investigated in research focusing on technology use in early education. We then go on to section 3 to explain the methods used in our own study followed by section 4 describing the toys investigated. In section 5, we present and evaluate the results of our study with three perspectives on play with smart toys. We sum up by presenting a framework introducing the game-based and toy-based play experiences potentials of smart and connected toys and end the paper with sections 6 and 7, discussing the results and proposing ideas for future research.

II. RELATED WORK

Attitudes of preschool educators towards the use of information and communication technology (ICT) are currently a major topic of research [10] [11] [12] [13].

One of the reasons for such interest lies in the fact that, as Hoffman, Park and Lin [10] suggest, today's question is how to assist kindergarten and preschool educators by providing knowledge, tools, and strategies required to respond to rapid changes in the learning environment, for example, how to make use of smart toys such as the current connected IoToys in playful learning. But, since there are no systematic studies of the IoToys suitability in the early childhood curriculum, some educators use ICT to a great extent while others do not use it at all [2] [12]. "While parental beliefs appear to play a significant role in children's development, play-learning beliefs remain relatively unexplored in the developmental literature", Fisher et al. write [14]. Consequently, it is of crucial importance for educators to become familiarized with the new tools that for example the IoToys addressed in this paper represent. For example, Chiong and Shuler (2010) discuss that adults keep young children motivated to use apps and smart toys by providing an extra prompt for the child to understand smart toys and games [15]. In other words, the child's capacities to use smart toys and games is influenced by the adult guidance. Thus, we claim that there is a need to recognize the importance of smart toys potentiality in early education to create opportunities for knowledge building and skills acquisition. The preschool teachers who participated in our study are using technology, such as tablets in learning activities. Their attitude toward digital devices and smart toys is positive. However, they are not familiar with smart toys' possibilities in an educational context.

In our earlier study (Ihamäki and Heljakka, 2018) we present the IoToys as a 21st century educational tool. Furthermore, we highlight the toy marketers' views of four IoToys' educational promises by describing the toys' possibilities to be used to offer rich, interactive, innovative and mobile learning experiences to preschool children both in the educational environment as well as in play during leisure time, as suggested by the designers and marketers of the smart toys [2].

Media technology, smart toys, and Internet content have an important role in the everyday life of young children [16]. Skills for understanding and using media are typically learned at home and from peers during the first years of a child's life. Therefore, children learn skills for using media outside of formal education and if granted opportunities, could also use this competence in their early childhood education [17]. For example, in Greece [13] they suggest that in order to successfully integrate ICT in educators' daily practices, it needs to be perceived as a mode of learning that should be embedded in the curriculum.

As the toys may be played with both in preschool as well as in the home, it is also important for the parents of young children to understand their potential. In this study, we try to increase knowledge on how it is important for educators and parents to understand the IoToys as a phenomenon, its products and potentialities, especially from the viewpoint of early learning. What needs to be critically evaluated, then, are the toy marketers' opinion on the capacities of these smart toys and the educational promises used in marketing them and what benefits the toys truly offer to children in terms of *edutainment*, that is, entertainment and education. Further, it is important for preschool teachers and parents to understand what kind of play experiences these smart toys can bring to the preschool context in the name of playful learning.

The many benefits of playful learning have been presented by Elkind (2007), who states that free, self- initiated, and spontaneous play contributes to the child's healthy, emotional and social development. Elkind stresses that a playful childhood, in fact, is the most basic right for children [18]. As we see it, educators need to remember to think about how their attitudes and opinions influence the employment of new tools such as toys that are available to offer children edutainment by learning through play. It is probable that the educators who have more experience with digital devices and who use digital devices personally, have more positive opinions generally. The educators interviewed for our study perceive digital technology (like iPads) as an useful educational tool that can help children to develop practical competencies and social practices. They also expressed intentions to use new technology in their kindergarten. Finally, it is also important to understand how parents' opinions and attitudes are formed, and how they may have an effect of making these kind of toys available for their young children in the context of home. At the same time, not every parent may have the possibility to offer children the opportunity to play with smart toys at home. However, since some IoToys are already available on the marketplace and their marketers are constantly trying to convince their buyers of their educational power, it is valid to ask whether or not they fulfil these promises for learning. Therefore, it is important to engage both educators and parents in a discussion around IoToys in order to know how they can be used in both unstructured, (toy-like) free play, and structured, (game-like) goal-oriented play.

III. Method

Häkkinen et al. (2003) suggested a multi-method approach that is process-oriented and takes into account different contextual aspects [19]. Our case study uses this approach in order to provide a holistic and complimentary description of the smart toys' possibilities for playful learning. Our study builds on the knowledge from the previous stages of our research: By using the findings of our study with preschoolers interested in the various play patterns IoToys as contemporary smart toys afford [3], and asking how this information from the play tests carried out with the IoToys correlates with smart toy marketers', preschool teachers', and parents of preschoolers' perspectives on play with digitally-enhanced toys, we set out to formulate a comprehensive understanding of how different stakeholder's views on smart toys address affordances related to playful learning. In other words, we analyze smart toy marketers' envisioned play experiences of three IoToys and compare these with preschool teachers' and parents' general attitudes towards smart toys and their educational potential. We examine the following research questions:

RQ1 (targeting the toy marketers): What kind of promises related to play experiences are the toy marketers giving for the IoToys, as reflected in their marketing texts (i.e. on websites and other advertising materials)?

RQ2 (targeting the preschool teachers): What kind of opinions do the preschool teachers' have regarding the play experiences related to smart toys used for playful learning in the preschool environment?

RQ3 (targeting the parents). What are the parents' opinions and observations of their children's play experiences with toys with digital dimensions, such as smart toys?

In order to analyze the viewpoints of toy marketers, preschool teachers and parents of preschool children who participated in the playtests in an earlier study [2] [3], we have collected three kinds of research materials: First, toy marketer's descriptions on three IoToys through websites, digital and printed marketing materials, second, a semi-structured survey for preschool teachers and third, a semi-structured survey for the parents of the preschool children.

The toys under investigation fulfil the criteria of IoToys, they are 'smart', and their connectivity usually occurs through mobile devices or some cases, smart toys contain their own computer (e.g., the CogniToys Dino and Fisher-Price's Smart Toy Bear.) The toys were chosen based on their age-appropriateness, gender-neutrality as character types of toys, and their availability on Amazon US (in August 2017). Also one reason for selecting these IoToys was the awards they have received. For example, the CogniToys Dino has received the Silver Honor 2016 Parent's choice award.

To analyze the research materials, we have used content analysis. The goal of content analysis is to provide knowledge and understanding of the phenomenon under study [20]. It provides the researcher with the possibility to make a close reading of the data through a systematic classification process of coding and identifying themes and patterns. Researchers immerse themselves in the data to allow new insights to emerge [21]. The method is also described as inductive category development [22].

The motivation to use three sets of research materials enriches the holistic view of adult perspectives on smart toys: The marketing materials help us to understand aspects of learning and contextualize and elucidate marketer motives in relation to the IoToys promises for learning. The preschool teachers' interviews inform us about the constraints and possibilities for using smart toys, such as IoToys in the context of early learning. Further, these interviews enable the articulation of aspects of learning with smart toys and help us to contextualize and elucidate individual attitudes and behavior, based on personal motives and perceptions in relation to use smart toys in the early education curriculum. We pursued those situational and contextual aspects of Finnish preschool that were identified by participants during our interviews. By using a semi-structured, thematic survey to explore parental attitudes, opinions and earlier experience of smart toys, we were able to draw parallels between ideas on toys' capacities to 'teach' at home. Next, we describe the rich material in our case study to illustrate the possibilities for playful learning with three contemporary smart toys.

A. Educational promises of the smart toys in our study: The marketers' perspective

We have chosen three smart toys (that also fulfil the criteria for IoToys), for which we have collected marketing materials to analyze the marketers' perspective of their educational affordances. All of these smart toys connect to the Internet via Wi-Fi. The connected toys employed in this study are *hybrid playthings*, which means that they are both physical artefacts and objects, which also function as portals to digital devices and socially shared content [3].

Our study employs toys that represent 'edutainment', and that according to their marketers, cater both to enjoyment and opportunities for learning: CogniToys Dino, Wonder Workshop's Dash and Fisher-Price Smart Toy Bear. These smart toys are briefly described in the following, based on their marketing texts:

CogniToys Dino

Amazon.com describes the CogniToys (by Elemental Path) as an educational toy, which includes stories, games, jokes and fun facts. Educational subjects including vocabulary, math, geography, science, which make children to engage more within an educational play based on their academic needs. The CogniToys Dino will constantly evolve because it is a cloud-connected, Wi-Fi enabled character allowing for the play experience to repeatedly improve and update automatically. The toy is said to engage children with a wide variety of content by encouraging learning and play using interactive dialogue. The CogniToys offers "FUNdemental Part of the CogniToys Experience", where each Dino comes with "a variety of custom modules to engage children in educational play including problem-solving challenges, geography games, historical fun facts and more" [23]. According to a marketing text published by Kickstarter, the CogniToys Dino "creates an atmosphere of fun, playful education! Built into the play experience is a number of custom modules that engage the child in educational play. As the interaction increases so will the challenge of the educational content continuing to become more challenging as the child learns" [24].

Wonder Workshop's Dash

Wonder Workshops' Dash's own website includes an educational part, on which educational benefits for using for Wonder Workshop's Dash in educational programs are introduced. The website describes how the Dash robot will "grow with students and it is easy for beginners to get started and then scaffolds learning through a meaningful curriculum." It is supportive of 21st century skills, by "encouraging students to continue building critical thinking, creativity, communication, and collaboration skills for tomorrow's job market" [25]. As the website for Wonder Workshops' Dash mentions, "play is a powerful teaching tool". Dash has its own comprehensive curricular resources for teachers to help students' practice computational thinking [26].

While the company sells Dash directly to families, it has also seen Dash and sidekick toy, Dot, to become part of schools' curriculum and coding clubs over the years. According to Kolodny, some 8,500 schools are using Dash and Dot around the world today [27].

Fisher-Price's Smart Toy Bear

Fisher Price's Smart Toy Bear is described as "an interactive learning friend as unique as your child" that comes without a screen. The toy responds to what the player says, and remembers things. It takes cues from its player, then invites play, talk, movement, imagination and learning. The Smart Toy is also able to recognize images with 9 'Smart Cards' included, which the player can choose to play activities like listen to stories, play games and 'go on adventures'. As the Smart Toy is constantly updated it is said to encourage social-emotional development, imagination and creativity: "Your Smart Toy learn new activities every month" [28].

A. Preschool teacher survey interviews

The importance of digital technology competencies of early education teachers is publicly recognized and supported in most countries around the world. One study from Mainland China showed that educators' educational level and ICT-related training were found most important in determining whatever they use digital technology in their teaching [12]. According to Konca et al. it is important to support educators' positive attitudes towards digital technology, but also their suggestion on using digital technology in kindergarten [11]. The research that investigated intentions of 50 early childhood educated teachers from University of Athens, Greece, using digital game-based learning, revealed that kindergarten educators in Greece generally have very positive views about the digital technology use and who owned a computer at home had more positive opinions. These educators perceive digital games as a useful educational tool for kindergarten environment [13].

In this study, we have used a semi-structured, thematic survey to explore two preschool teachers' views on smart toys and their use in the preschool context. The teachers have informed that the children in their group each have their personal tablet at preschool, which they are allowed to use in supervision for a limited time per day. Out of the 20 preschool-aged children (5-6 years of age), who took part in our play tests in previous stages of our research, we have asked how many of them have a mobile phone of their own. Of the children that participated in our study, 10 reported owning a mobile phone. The teachers were asked about the use of digital media and toys in the preschool context in general, their ideas on smart toys, and their opinion regarding the affordances of these toys that relate to playful learning.

B. Parental survey interviews

Previous studies have demonstrated that the extent to which parents' guide their children's media use, digital games and toys, and the strategies they apply, are related to demographic variables, such as the parent's age, gender, and education or income level. In addition, the parent's own media use and skills, and family context variables, such as family size, marital status and the number of media screens at home are important as well [29]. Furthermore, parents who are less skilled in using media and digital devices themselves may find it more difficult to install parental controls on the devices, games and smart toys or to discuss the media content critically with their children as compared to media literate parents [30]. Indeed, researchers have shown that parents' mediation strategies are in accordance with their views on various effects of media content on children. Parents who feel that media offers educational or entertainment opportunities more often co-use the media, digital devices and smart toys with their child, or actively discuss the content [31]. Many parents feel that through media use their young children develop their physical, cognitive and emotional capacities, a wide range of media skills, defined as the child's knowledge and understanding of the role of media and technology in society [32].

In our study, we have conducted survey interviews concerned with parents' perspectives on digitally enhanced toys such as smart toys on a general level, the play of their children with these toys, and these toys' capacity to cater for playful learning. We have used a semi-structured, thematic survey to explore parental attitudes, opinions and earlier experiences of smart toys. We have used a thematic survey to explore parental and kindergarten teachers' attitudes and parent's experiences of digitally enhanced toys in general. Thematic survey focuses on mainly on investigating the following questions:

- Does your child play with the (digitally-enhanced) toy alone or in the company of other children?
- Do you think that playing with this kind of a toy teaches the child new skills?
- Does the child simultaneously use (other) mobile devices when playing with the toy?
- Does your child play with the toy in any of the following ways: nurses the toy; uses the toy in narrative play (gives the toy a role and lines of speech in play); explores the toys' mechanical features; tries to teach the toy new skills; uses the toy as a bedtime companion?

We have recruited the parents of 20 preschool-aged children (altogether 14 parents, both male and female), the survey interviews with whom the interviews were carried out in October 2017 [2] [3].

The parents' were asked about their attitudes to children's use of digital media and toys in general, their opinions on digitally-enhanced toys such as smart toys, and their observations on their children's play with smart toys related to solitary and social play, learning experiences, use of mobile devices as extensions of toy play, play patterns demonstrated by their children, and the importance of the toys' various playful affordances.

IV. RESULTS

Next, we report the results of our study, organized around three perspectives on play with smart toys: the promises of playful learning as described by the smart toy marketers, the preschool teachers' attitudes and opinions on smart toys, and the parental perspective concerning this type of toys.

Toy marketers' perspective on smart toys

We begin by highlighting several aspects of marketers' ideas on smart toys and their suitability to be used in playful learning, which helps to set the context of this study. We concentrate on three IoToys, which represent the most recent evolution of smart toys. These smart toys have won several awards and are named 'children's best toys' [23]. In their communication marketers, have used words like 'future technology', 'future education' and they have also given the picture that these are 'dream toys for children' with which they also can learn for example languages, pronouncing skills and mathematical skills. These marketing texts are also directed towards parents' in the extent to which they give a promise of learning opportunities to children. The marketers also use examples of how many kindergartens already have used for example the Dash robot in kindergarten.

'Educational value' is frequently is used as a marketing concept for the connected toys of the present. In particular, the usage of the word "smart" occurs in high frequency to describe the level of intelligence in the toy that will transfer to its user. For example, on the website for the CogniToys Dino it is explained that that the toy could teach mathematics like addition, subtraction and division. The CogniToys Dino also offers digital games, with educational potential, like 'Math Mouse', 'Capital Quest', 'Country Quest', and so on [33]. The marketers describe the smart toys to learn and grow alongside children. Further, they believe that the toys can be used as educational tools for example in kindergarten. For instance, the toy marketer for Wonder Workshop's Dash claims that their products and apps may teach students from kindergarten and up how to code. They offer six apps; students can for example tackle a series of challenges and in- app puzzles that introduce the concept of coding. The marketer describes the Wonder App 'to use machine programming and an original visual design to advance young students' understanding of computer science'. Again, the Blockly app is a block-based coding app that introduces children to programming [27].

The Fisher-Price Smart Toy Bear is marketed as an educational toy, which children can learn important concepts with through engaging in interactive play. Children can train their smart toy to recognize their voice through embedded voice recognition features, or to respond to their commands with a selected option. For example, the developers claim that the smart toy adapts to developmental changes while remembering the child's name, favorite colors', foods and more. The toy suggests adventures that they should 'go on adventures together', along with stories and games [29].

In sum, as smart toys, the IoToys chosen for our study have, by their marketers, been given promises regarding various play experiences that are believed to have values for use in education [2] [3]. From the perspective of the Playful Experience Framework (or PLEX), the study demonstrated, how the toy marketers' ideas on the IoToys on the one hand correlated with the experiences categorized in the PLEX framework, and thus in both areas of play experiences: game- based and toy-based play experiences as categorized by the authors (See Table 1). The five most frequently occurring play experiences communicated in the analyzed marketing texts (websites, packaging), are listed in the following:

1. Challenge (game-based PLEX); Content related to educational subjects, coding and games
2. Humor, relaxation, thrill (toy-based PLEX); Storytelling, jokes
3. Fellowship (toy-based PLEX); Collaboration
4. Exploration (toy-based PLEX) 'Go on adventures'
5. Discovery (toy-based PLEX); Content related to play by image recognition through 'SmartCards'

However, some differences could be detected in the envisioned play experiences of the IoToys: For example, the CogniToys Dino and Wonder Workshop's Dash included more game-based affordances for playful learning (challenge, competition, completion and control), whereas the Fisher- Price's Smart Toy Bear seemed to afford more toy-based play experiences for playful learning (for example, through humor, relaxation and sympathy).

Preschool teachers' perspective on smart toys

Hannaway and Steyn (2017) propose that teachers' use of digital technology influence their pedagogy and the ways knowledge is created in the classroom [34]. Again, teachers' attitudes toward digital technology use of young children are informed by the children's age: Mertala (2017) studied Finnish trainee teachers' perceptions of the role of digital technology in early year education and concluded that technology-related concerns for the youngest children were related to the children's physical health, whereas for older children it was their intellectual health [35].

The interaction with smart toys is generally not seen as very appealing from an educational perspective [36]. According to Ruckenstein (2010), the Nordic discourse on digital toys often constructs these toys as unnatural intrusions into the lives of small children [37]. As the category of smart toys has evolved and developed rapidly into more pedagogically informed devices, we were interested in preschool teachers' attitudes and opinions on these digitally-enhanced toys. The preschool educators' (n=2) perspective in this study allowed us to consider the differences that formal and informal learning environments may provide to the connected toys' usage as playthings. The teachers' informed us that the preschool currently uses learning games, but limits their use to a certain time of the day. The two female educators interviewed for our study identified toys such as Storyteller Yano and Tamagotchi to represent toys with digital enhancements and dimensions. Both interviewees considered the educational affordances of the smart toys to be highly important. Out of the play patterns described in our survey the teachers listed the narrativizing of the toy (i.e. storytelling with it) to be the most important. According to the research materials collected from the surveys, the teachers' attitude towards educational technology was positive, as they claimed that "technology could be used even more in preschool". However, a comment made by one of the preschool teachers supports Ruckenstein's thinking of digital toys as 'intrusions' in the early education context [37]: "I see that digital toys are more suitable in the domestic environment than in the kindergarten" (preschool educator, 29 years in the profession). From the perspective of the PLEX Framework, preschool teachers considered *control* ("the player is able to teach the toy new skills") as a game-based experience related to smart toys to be important for playful learning. Again, the toy- based play experiences of *exploration, expression, nurture* and *relaxation* were marked as important facets of smart toys' dimensions in terms of playful learning. Together, the five most prominent play experiences were listed in the following order of importance in reference to playful learning by the teachers:

1. Expression (toy-based PLEX); narrativizing theory
2. Nurture (toy-based PLEX); nurturing of the toy
3. Control (game-based PLEX); teaching the toy new skills
4. Exploration (toy-based PLEX); exploring the toy's mechanical features
5. Relaxation (toy-based PLEX); using the toy as a bedtime companion)

Parents' perspective on smart toys

We have used a thematic survey to explore parental attitudes and experiences of digitally enhanced toys. Although the 14 parents of the 20 preschoolers who participated in our play tests were interviewed about a wide range of topics in relation to digitally enhanced toys in general (to be presented in the subsequent stages of our research), this study focuses mainly on investigating the following questions:

- Do you think that playing with a digitally-enhanced (smart) toy teaches the child new skills?
- Does your child play with the toy in any of the following ways: Nurtures the toy; uses the toy in narrative play (gives the toy a role and lines of speech in play); explores the toys' mechanical features; tries to teach the toy new skills; uses the toy as a bedtime companion?

Altogether, 14 parents (n=11 female, n=3 male) from different socio-demographic backgrounds participated in our semi-structured survey. Ten of the parents reported their child as owning some kind of toy with digital dimensions. Half of the parents considered their child as learning something while playing with digitally enhanced toys. The most popular play pattern the parents reported their child as carrying out with the toy was exploring the toys' mechanical features (9 children), and the second most popular play pattern was to use the toy in narrative play (7 children). The third most popular pattern on the list of play patterns children carry out with toys with digital dimensions was to nurture the toy and use it as a bedtime companion (6 children).

Considering the educational features of the digitally enhanced toys, the parents who answered these questions reported the most important feature as the toys' capability of teaching the child to how to count and how to be self-expressive, teaching good manners, and taking other players into consideration. The toys' ability to teach their players to read, ask questions, and be self-expressive (e.g., to come up with stories) were considered somewhat important by the parents. The majority of parents (8/14) answered that the toy's educational aspect is its most important aspect. We consider this to represent the toy-based play experience of *challenge*. Together, the five most prominent play experiences were listed in the following order of importance in reference to the perspective of parents to playful learning:

1. Challenge 8/14 (game-based PLEX)
2. Expression, exploration 7/14 (toy-based PLEX)
3. Relaxation 6/14 (toy-based PLEX)
4. Fellowship 3/14 (toy-based PLEX)
5. Nurture 2/14 (toy-based PLEX)

V. EVALUATION

In this study, we have explored play experiences related to playful learning experiences of smart toys from the marketers, teachers and parents' perspectives, which we analyzed through the PLEX Framework. We suggest that these play experiences can best be evaluated from the perspective of using smart toys in early education, and by dividing them further into game-based and toy-based learning experiences, which suit different learning goals. We have found that from the perspective of toy marketers', play experiences that are connected with contemporary smart toys, or IoToys, relate most to the play experiences of *challenge*, which we consider an experience best understood as a play experience for game-based learning. In other words, toy marketers suggest that activities such as learning of educational subjects, coding and playing of games are meaningful for the players. Other play experiences envisioned by the toy marketers are the experiences related to storytelling and telling of jokes, which we interpret as the toy- based play experiences of humor, relaxation and thrill. Other play experiences accentuated in the toy marketers' perspective are the play experiences of fellowship, exploration and discovery, which relate to the IoToys under scrutiny in terms of their affordances of pursuing collaboration between players, 'going on adventures' with the smart toy to explore new content communicated by the toy, and finally, enable the player to discover new play ideas by triggering interaction between the toy and the player by the use of image recognition through 'smart cards'.

According to preschool teachers *control* ("the player is able to teach the toy new skills") as a game-based experience related to smart toys and the toy-based play experiences of *exploration, expression, nurture* and *relaxation* were marked as important facets of smart toys' dimensions in terms of playful learning. Finally, the parents listed the toys ability to educate—the game-based play experience of *challenge*, to be a key aspect of smart toys in terms of playful learning. The toy-based play experiences of *exploration, expression, nurture* and *relaxation* were marked as important facets of smart toys' dimensions in terms of playful learning. For all detected play experiences of smart toys, see the following Table 1.

TABLE I. THE GAME-BASED AND TOY-BASED PLAY EXPERIENCES POTENTIALS OF SMART AND CONNECTED TOYS, PARENTS' AND TEACHERS' PERSPECTIVES, TOY MARKETERS' PERSPECTIVES, AND RESULTS OF THE PLAY TESTS IN OUR EARLIER STUDY.

	Playful Experience	Parental and preschool teachers' perspectives on Play Experiences	Toy marketers' perspectives on Play Experiences	Results of play-tests in our study
Play experiences for gameplay and game-based learning	Challenge	Importance for the parent that the toy is educational: 8 parents out of 14 parents most important PLEX 2 parents out of 14 parents second important PLEX 1 parents out of 14 parents third important PLEX	"CogniToys offers game challenges like, "Dog Quiz game, Math Mouse, Ocean Quiz, Animal Expert, Capital Quest" etc." (www.CogniToys.com)	Children's abilities are tested by the IoToys' demanding tasks. (Wonder Workshop's Dash)
	Competition	Parents or preschool teachers did not mention this PLEX in our survey.	"Wonder Workshop's Dash has League Robotics Competitions to build teamwork and community spirit" (www.makewonder.com)	Children can contest with their earlier experiences with IoToys. (Wonder Workshop's Dash)
	Completion	Parents or preschool teachers did not mention this PLEX in our study.	"Program Wonder Workshop's Dash to be a musical performer with Xylo. Used with the Xylophone accessory, kids can program Dash to play their favourite song or make new tunes of their own." (www.makewonder.com)	Finishing a major task, like listening to the IoToys' story. (Fisher-Price Smart Toy Bear, CogniToy Dino)
	Control	Play experience detected by the parents: The child aims to teach the toy new skills, 4 parents out of 14 parents reported this PLEX. Teachers evaluated this to be the 3rd most important PLEX of all.	"CogniToys offers General Command Prompts: Let's play a game, tell me a story, tell me a joke, Play a Song etc." (www.CogniToys.com)	Commanding IoToys with an IPad. (Wonder Workshop's Dash, Fisher-Price Smart Toy Bear, CogniToy Dino)
Play experiences for open-ended play and toy based-learning	Discovery	Parents or preschool teachers did not mention this PLEX in our study.	"Wonder Workshop's Dash has more than 600 built-in-tutorials, challenges and projects, the apps let children explore programming at their own pace." (Wonder Workshop's Dash Robot – Apple)	Children's imaginative play with IoToys presents a way of playing what the designer never even thought, like using the IoToys as a lamp. (Wonder Workshop's Dash)
	Exploration	Play experience detected by the parents: The child explores the mechanical features of the toy: 7 parents out of 14 parents reported this. Teachers evaluated this to be the 4th most important PLEX.	"CogniToys update automatically as new content becomes available." (www.CogniToys.com)	Investigating an object or situation with the IoToys. (Wonder Workshop's Dash, Fisher-Price's Smart Toy Bear, CogniToys Dino)
	Expression	Play experience detected by the parents: The child narrates the toy: 8 parents out of 14 parents reported this Teachers evaluated this to be the 1st most important play experience.	CogniToys expresses itself like green light in his/her mouth means "I am ready to play", blue lights mean I am talking, yellow light means I am listening and so on". (https://vimeo.com/158246191)	Children make creative things by coding. (Wonder Workshop's Dash)
	Fantasy	Play experience detected by the parents: the child anthropomorphizes the toy: 2 parents out of 14 parents reported this PLEX.	"Wonder Workshop's Dash makes robotics as delightful as finger painting. A picture-based coding language built for kids, children create detailed behaviours for Dash, creating the robot pet, pal or sidekick of their dreams." (www.makewonder.com)	An imagined experience, e.g. the "IoToy can teach me to fly". (Fisher-Price's Smart Toy Bear, CogniToys Dino)
	Fellowship	Importance for the parent that the toy enables social play: 3 parents out of the 14 parents most important PLEX 7 parents out of the 14 parents second important PLEX 1 parents out of the 14 parents third important PLEX	"The Fisher-Price Smart Toy Bear is an interactive learning friend with all the brains of a computer, without the screen! The more your child plays with Smart Toy, the more this remarkable furry friend adapts to create personalized adventures the two of them will love! Sounds like the start of a true friendship!" (www.momvstheboys.com 2015)	IoToys like the Dash Robot has their own community to share experiences of one's own toy with others. (Wonder Workshop's Dash)
	Humour	Parents or preschool teachers did not mention this PLEX in our study.	"The CogniToys Dino tells jokes, for example: "Punny yodeling joke: "Knock knock." "Who's there?" "Little old lady." "Little old lady, who?" "Wow! I didn't know you could yodel!" (www.pinterest.co.uk/cognitoys/)	IoToys give children fun and joyous experiences by telling children stories and jokes. (Fisher-Price's Smart Toy Bear, CogniToys Dino)
	Nurture	Importance for the parent that child nurtures the toy: 2 parents out of the 14 parents most important PLEX 1 parents out of the 14 parents second important PLEX 9 parents out of the 14 parents third important PLEX Teachers evaluated this to be the 2nd most important PLEX	"Fisher-Price's Smart Toy Bear of content really allows children to communicate with the Smart toy, which offer unique personalized content and nurturing them through play". (www.smartoy.com)	Children want to take care of their IoToys. (Fisher-Price's Smart Toy Bear)
	Relaxation	The child uses the toy as a bedtime companion: 6 parents out of the 14 parents reported this. Teachers evaluated this to be the 5th important PLEX	"Fisher-Price's Smart Toy bear climb under the covers and enjoy bedtime stories, soothing music or nature sounds to help wind down after a long day". (www.smartoy.com)	Children comment that the IoToy can read them a bedtime story. (Fisher-Price's Smart Toy Bear, CogniToys Dino)
	Sensation	Parents or preschool teachers did not mention this PLEX in our study.	"CogniToys are smart devices in toy form, tailored just for kids to provide an educational and entertaining experience without the need for a screen". (www.CogniToys.com)	Children think that IoToys are exciting as they stimulate the senses and give children feedback. (Fisher-Price's Smart Toy Bear, CogniToys Dino)
	Sympathy	Parents or preschool teachers did not mention this PLEX in our study.	"Fisher-Price's Smart Toy Bear: it's never-ending fun that encourages social-emotional development, imagination and creativity with your child". (www.smartoy.com)	Children can share emotional states with their IoToys. (Fisher-Price's Smart Toy Bear, CogniToys Dino)
	Thrill	Parents or preschool teachers did not mention this PLEX in our study.	"Fisher-Price's Smart Toy Bear surprise means to explore a mysterious cave, soar on a magic carpet, rescue a pod of dolphins and more adventures." (www.smartoy.com)	Children's excitement derives from to taking risks with the IoToy, e.g. by listening to a ghost story told by the toy or risk-taking in reference to coding with the IoToys. (Fisher-Price's Smart Toy Bear, CogniToys Dino)

VI. Discussion

The IoToys have been on a high growth in numbers for some time now. Simultaneously, parents want to take a more active role in their children's education and give them tools for that. The IoToys as an emerging category of smart toys, seems to deliver both entertainment and education, offer opportunities for free play and connect children to play games, and make activities by learning new skills, like mathematics and languages. Toy companies work to keep children and parents engaged, offering new ways to both play and learn. Playing with contemporary smart toys connected to the Internet, future play can be shaped as children may connect with others, in the online and offline domains, public and private spaces and in formal and informal learning environments. In other words, children's playing is no longer confined to the physical space but extends to virtual realms, which means that the IoToys offer opportunities to engage in the kinds of transmedia and trans-domain practices. These practices transform the toy to operate as a kind of boundary object [38], enabling the children to transverse different domains and practices with ease. For example, the Fisher-Price Smart Toy Bear learns how the child plays and recommends new activities because of its connection to the Internet through Wi-Fi.

In our study, we have presented three different perspectives on play with smart toys: Toy marketers' suggestions on the IoToys' potential for playful learning, preschool teachers' views on smart toys suitability to be used in the early learning context, and parents' attitudes towards smart toys based on their observations of children's play in the domestic sphere. All these stakeholders have understood that play evolves, and modern technology can help a child feel more empowered, capable and they can learn with toys. Still the question of how to make use of these contemporary smart toys in early education can be raised. The preschool teachers' attitudes will influence their teaching activities when using digital devices, games and smart toys in their education. Clearly, more research is needed to understand how preschool teachers could use for example the IoToys for learning. It is also important to remember that these toys function as entertainment, and children playing with the smart toys in home environments may not understand all their educational features. Like many studies have shown, when parents are more involved with their children's education they also want toys to teach something to their children. Therefore, the parental perspective on what is considered important in terms of play experiences that may contribute to playful learning should be studied and acknowledged even more.

VII. Conclusion and Future Research

The results of our study present that marketers have seen the educational potential of contemporary smart toys, andare using that in their communication. Moreover, they also reference early education parties, which have used their smart toys as a part education.

Despite their complex technological systems, smart toys designed for the mass market come in relatively affordable prices for consumers of the Western world and are in this way more likely to enter the domestic playscapes of young children. Although, at many times, digitally-enhanced toys are still considered to fit informal learning environments such as the home better than the formal educational context of preschool, we believe that the ongoing digitalization development will eventually also introduce more preschools with more connected toys. We believe, that contemporary smart toys are bridging play and learning in positive ways, and could be used even more as a part of informal and formal learning. Therefore, it is beneficial for both educators and parents of young children to know the various possibilities these playthings afford both in terms of different play experiences. In future stages of our research, our aim is to conduct a longitudinal study with both preschoolers, their parents and their teachers in order to find out more about the play experiences the contemporary smart toys afford.

In our study we explored smart toys' potential to deliver experiences related to playful learning. One key aspect of toys such as the CogniToys Dino, Fisher-Price's Smart Toy Bear and Wonder Workshop's Dash Robot are their game-based and toy-based features and functions, which are suggested to have educational outcomes when used in play.

We hope this paper to have highlighted the three perspectives on play with smart toys from the viewpoints of toy marketers', preschool teachers and parents. As our results have shown, preschool teachers are generally positive towards the IoToys, but are not necessary sure about how to use them in education. This means that early educational professionals must be more informed about new technologies affordances for playful learning. Furthermore, as our findings show, parents are generally welcoming to smart toys into their children's lives as well, because they offer new modes of playing and learning, yet do not really know their potential. These modes forplayful learning could therefore be evaluated based on play experiences related to smart toys' as game-based and toy-based affordances, as presented in our paper. In this way, their suitability could be better considered to be used for playful learning in different learning situations and contexts.

Acknowledgments

We wish to express our gratitude to the parents and preschoolteachers for participating in our surveys. This study has been partially funded by the Centre of Excellence in Game Culture Studies (decision #312396).

References

[1] A. Lucero, E. Karapanos, J. Arrasvuori, and H. Korhonen," Playful or Gameful? Creating Delightful User Experiences", in Interaction Vol. 21, No. 3, pp. 34-39, 2014.

[2] P. Ihamäki and K. Heljakka, "Smart, skilled and connected in the 21st century: Educational promises of the internet of toys (IoToys)," in Proceedings of the 2018 Hawaii University International Conferences, Arts, Humanities, Social Sciences & Education, Prince Waikiki Hotel, Honolulu, Hawaii, January 3–6, 2018.

[3] K. Heljakka and P. Ihamäki, "Preschoolers learning with the Internet of Toys: From toy-based edutainment to transmedia literacy", in Seminar.net, International Journal of Media, Technology & Lifelong Learning, Vol. 14, No. 1, pp. 85–102, 2018,

[4] J. M. Healy, "Failure to connect: How computers affect our children's minds – For better and worse, New York: Simon & Schuster, 1998.

[5] Z. Corbyn "The future of smart toys and the battle for digital children", in The Guardian, 2016. [Online]. Available from: https://www.theguardian.com/technology/2016/sep/22/digital-children-smart-toys-technology

[6] Y. Ren, J. Shen, J. Wang, J., Han, and S. Lee, "Mutual Verifiable Provable Data Auditing in Public Cloud Storage", in Journal of Internet Tech, Vol. 16, No, 2, pp. 317-323, 2015.

[7] E. I. That, "Context data model for privacy," PRIME Standardization Workshop, IBM Zurich, 6 Pages, 2006.

[8] L. Rafferty, P. C. K, Hung, M. Fantinato, S. M. Peres, F. Iqbal, S-Y. Kuo, and S-C. Huang, "Towards a Privacy Rule Conceptual Model for Smart Toys", In Proceedings of the 50th Hawaii International Conference on System Sciences 2017, pp. 1226-1235, 2017, doi: 10.24251/HICSS.2017.146

[9] D. Holloway and L. Green, "The Internet of Toys," in Communication Research and Practice, Vol. 2, No. 4, pp. 506–519, 2016.

[10] E. S. Hoffman, E. Park, and M. G. Lin., "Beyond professional development: A case study of implementing iPads in early childhood education," in D. Slykhuis and G. Marks (Eds.), Proceedings of Society for Information Technology & Teacher Education International Conference, Chesapeak, VA: Association for the Advancement of Computing in Education (AACE), pp. 2008–2015, 2015.

[11] A. S. Konca, E. Ozel, and H. Zelyurt, "Attitudes of preschool teachers towards using information and communication technologies (ICT)", in International Journal of Research in Education and Science (IJRES), Vol. 2, No. 1, pp. 10–15, 2016.

[12] X. Liu and J. Pange, "Early childhood teachers' access to and use of ICT in teaching: The case of Mainland China", in Proceedings of Global Learn, Berlin, Germany: Association for the Advancement of Computing in Education (AACE), pp. 590–596, 2015.

[13] K. Nikolopoulou and V. Gialamas, "ICT and play in preschool: Early childhood teachers' beliefs and confidence", in International Journal of Early Years Education, Vo. 23, No. 4, pp. 409–425, 2015, doi: 10.1080/09669760.2015.1078727

[14] K. R. Fisher, K. Hirsh-Pasek, R. M. Glinkoff, and S.G. Gryfe, "Conceptual split? Parents' and experts' perceptions of play in the 21s century", in Journal of Applied Developmental Psychology, Vol. 29, No. 4., pp. 305–316, 2008.

[15] C. Chiong and C. Shuler, "Learning: Is there an app for that? Investigations of young children's usage and learning with mobile devices and apps", New York: The Joan Ganz Cooney Center at Sesame Workshop, 2010.

[16] M. Ito, "Hanging out, messing around, and geeking out: Kids living and learning with new media", MA: MIT Press, 2010.

[17] L. Plowman, C. Stephen, and J. McPake, "Growing up with technology, Young children learning in a digital world, NY: Routledge, 2010.

[18] D. Elkind, "The hurried child: growing up too fast too soon", In 25th Anniversary edition New York, Da Capo Lifelong Learning, pp. XVII, 2007.

[19] P. Häkkinen, S. Järvela, and K. Mäkitalo, "Sharing perspectives in virtual interaction: Review of methods of analysis". In (Eds.) B. Watson, S. Ludvigson and U. Hoppe, Designing for Change in Networked Learning. Proceedings of the International Conference on Computer Support Collaborative Learning, Dordrecht: Kuwer, pp. 395–404, 2003.

[20] B. Downe-Wambolt, "Content analysis: Method, applications and issues", in Health Care for Women International, Vol. 13, pp. 313-321, 1992.

[21] N. L. Kondraki and N. S. Wellman, "Content analysis: Review of methods and their applications in nutrition education", in Journal of Nutrition Education and Behavior, Vol. 34, pp. 224-230, 2002.

[22] P. Mayring, "Qualitative content analysis", Forum: Qualitative Social Research, Vol. 1., No., 2, [Online]. Available from: http://www.qualitative-research.net/fqs-texte/2-00/02-00mayring-e.htm

[23] CogniToys Dino, Powered by IBM Watson, Kids Cognitive Electronic Learning Toy, Amazon. [Online]. Available from: https://www.amazon.com/CogniToys-Powered-Cognitive-Electronic-Learning/dp/B014IO4HYS/ref=sr_1_1?ie=UTF8&qid=1534058082&sr=8-1&keywords=cognitoys+dino

[24] CogniToys: Internet-connected Smart Toys that Learn and Grow, Kickstarter. [Online]. Available from: https://www.kickstarter.com/projects/cognitoys/cognitoys-internet-connected-smart-toys-that-learn

[25] Wonder Workshop website, Education, [Online]. Available from: https://www.makewonder.com/education/

[26] Wonder Workshop website, Curriculum, [Online]. Available from: https://education.makewonder.com/curriculum/

[27] L. Kolody, "Kids can now program Dash and Dot robots through Swift Playgrounds", In TechCrunch.com 18.10.2016. [Online]. Available from: https://techcrunch.com/2016/10/18/kids-can-now-program-dash-and-dot-robots-through-swift-playgrounds/

[28] Fisher Price Smart Toy, [Online]. Available from: https://www.smarttoy.com/products

[29] D. Manessis, "Early childhood post-educated teachers' views and intentions about using digital games in the classroom", in Proceedings of the 5th European Conference on Games Based Learning, pp. 753-758, 2011.

[30] J. De Haan, "Late on the curve; cause and consequences of differences in digital skills", in E. Ferro, Y., Kumar Dwivedi, J.R. Gil-Garcia and M. D. Williams (Eds.) Handbook of research on overcoming digital divides: Constructing an equitable and competitive information society, pp. 292–308, Hershey: Information Science Reference, 2010.

[31] N. Sonck, P. Nikken, and J. De Haan, "Determinants of internet mediation: A comparison of the reports by parents and children" in Journal of Children and Media 7, pp. 96-113, 2013.

[32] J. Marsh, G. Brooks, J., Hughes, L., Ritchie, S. Roberts, and K. Wright, "Digital beginning: Young children's use of popular culture, media and new technologies". Sheffield: University of Sheffield, 2005. [Online]. Available from: https://www.researchgate.net/profile/Greg_Brooks/publication/265183910_Digital_beginnings_Young_children%27s_use_of_popular_culture_media_and_new_technologies/links/5473599e0cf2d67fc036d3df/Digital-beginnings-Young-childrens-use-of-popular-culture-media-and-new-technologies.pdf?origin=publication_detail

[33] Cognitoys, Dino Commands, 2018. [Online]. Available from: https://cognitoys.com/pages/dino-commands

[34] D. M. Hannaway and M. G. Steyn, "Teachers' experiences of technology-based teaching and learning in the foundation phase. In Early Child Development and Care, Vol 187, No. 11, pp. 1745-1759, 2017.

[35] P. Mertala "Digital technologies in early childhood education – a frame analysis of preservice teachers' perceptions', in Early Child Development and Care, pp. 1-14, 2017. doi: 10.1080/03004430.2017.1372756

[36] M. Ruckenstein, "Spatial extensions of childhood: from toy worlds to online communities", in Children's Geographies, Vol. 11, No. 4, pp. 476-489, 2013.

[37] M. Ruckenstein, "Toying with the world: children, virtualpets and the value of mobility", in Childhood, Vol. 17, No. 4, pp. 500-513, 2010.

[38] G.C. Bowker, S. Timmermans, A.E. Clarke, and E. Balka "Boundary objects and beyond: Working with Leigh Star", Cambridge, MA: The MIT Press, 2015.

A Player-centric Game Design Paradigm with Compassion

Game design challenges for players with Parkinson Disease

Aslihan Tece Bayrak
Dept. of Game Development
Media Design School, MDS
Auckland, New Zealand
tece.bayrak@mediadesignschool.com

Abstract—**The population playing games vary from young ages to older adults, spread across people with variations of physical and cognitive abilities, and with different physical and mental health conditions. When additional sub-demographics, such as elders, people with specific disabilities or chronic diseases, etc., are taken into account, the territory for accessibility in game design and accessibility related design challenges are underexplored. Although the emphasis on usability in game development somewhat follows common usability guidelines developed for human computer interaction, usability context for games and game-like experiences includes more than the interface or the interaction modality. This body of work presents a player-centric game design paradigm to allow a closer exploration for challenges of designing games for sensitive demographics, with a focus on Parkinson Disease (PD) patients. The contribution is not a clear-cut recipe but an initial model towards designing accessible games for sensitive demographics.**

Keywords-games; game design; player-centric design; accessibility; parkinson disease; games for health.

I. INTRODUCTION

Games have been receiving attention from health and well-being research due to their compelling nature. Over the years, literature on games for health has grown with both use of commercial games and development of custom made games for health interventions [2]-[5]. Although positive results and various key learnings are reported by many studies, clear contributions on reappropriation of existing games for specific purposes or developing games for people with chronic and/or neurological diseases is limited. Researchers have mostly adopted or applied existing game design models [9][19] or developed on an ad-hoc basis in order to create a suitable product without much emphasis on the design practice [4][5][14][15]. Therefore, findings from these studies are fragmented and hard to reconcile for further studies. Lack of a methodology that facilitates design exploration for these demographics seems to be a result of limited design research within games research [21]. Both for the use of existing commercial games and with the purpose of developing custom games, there is a need for further design research to ensure suitability and accessibility of games for special demographics [4][5]. Even for the application of accessibility guidelines, practice for the accessible games does not go much further than the user interface or interaction [16]. A suitable interaction modality and a satisfying user experience are integral to the creation of engaging and enjoyable game experiences. Furthermore, it is essential for the player to be able to see past the interaction modalities while playing a game so that they could fully engage with the activities and events in the game. This paper aims to encourage exploration of these issues by introducing a new design paradigm that helps the researchers to focus on player and player's perception of the system (the game). Games for health research hopes to leverage the compelling nature of games for positive health outcomes; therefore, games that are designed for this purpose need to be player-centred and favouring player's conditions.

Moving forward from the existing research, this paper presents a game design paradigm to explore designing for a set of impairments identified for a special player group—Parkinson Disease (PD) patients. Even though these challenges seem to be specific for PD, the model developed to explore player-centric game design remains relevant regardless of the specificity of the target player group. The model promotes discussion to understand the relation between players' conditions and the layers of the game. For this purpose, the paper is composed of following sections. Section II presents the motivation with a discussion on the challenges of game design for PD patients. Section III explains player-centric design and presents the player-centric design model with further elaboration on the main elements of the model. Section IV ends the paper with a conclusion while also pointing to future work.

II. MOTIVATION FOR DESIGN RESEARCH

Nielsen and Norman group defines usability as "a quality attribute that assesses how easy user interfaces are to use" [6]. A usable interface should be easy for a user to become familiar with and competent in; to achieve their objective; to discover previously less known attributes when need arises; to recover from an error; and to recall how to use the interface on subsequent visits. Accessibility is also well defined within user experience domain and has been a core element of usability evaluations. Similarly, accessibility of a game refers to the ability to enable the participation of people with disabilities to interact with the game and play without feeling any barriers during their interaction with the game or during overcoming the challenges presented by the game. Extending from usability research, playability and player-game interaction research has finally started gaining some speed to explore practical applications for game usability challenges [16]. The interest in the field includes all facets of player experience including but not limited to engagement, immersion, etc.

A set of guidelines was collated to inform developers on the special constraints for their design [11][14][17][18] while also pointing out the lack of further development in the game design discipline. Game design in its practice is agile and opportunistic, yet the design practice itself is under-researched. A study by Isbister and Mueller on variety of interaction modalities especially in the field of movement based games is one of the few that evaluates conditions of interaction and suggests strategies for a more successful design of movement-based systems, mainly games [17]. The guidelines provide insights for the design of interaction modalities, necessary feedback compatible with the interaction device and activity, extends of self-expression, challenges and fun. Among the few studies that attempt to develop games for PD patients, Assad et. al. previously suggested design principles for designing games for PD, particularly for motion based games for rehabilitation purposes [11]. Although informative, the principles are prescriptive for a specific type of game rather than allowing a wider applicability, and the paper does not present a clear methodology for design. Another model named as "extended model" by Gerling et. al [18] inspired the development of the paradigm presented in this paper for two reasons: the model presented in the paper is restricted with the structural constraints of existing models and it is limited in terms of acknowledging the role of the player from a player experience perspective even though there are discussion on accessibility. Nevertheless, the notes on the player's abilities (both cognitive and motor abilities, such as attention span, short-term memory, repetitive input, etc.) to be considered as resources resonates with the perspective of this paper.

A. On Challenges of Game Design for PD Patients

Citing form Sutton-Smith, Zimmermann and Salen [13] refers to game experience as a combination of five dimensions: visual scanning, auditory discriminations, motor responses, concentration and perceptual patterns of learning. Players scan the entire scene based on the visual and auditory signs while concentrating on events and signals provided by the game. They perform actions based on the demands of the game and proceed whilst scanning for visual and auditory cues. This cycle continues as the player carries on playing the game. As they do so, players learn more about the patterns of play and improve their standing against the game from familiarity to higher expertise. At its core, the cycle of play stands on the perceptual understanding of the game world, the processing time of the perceptions, and the response from the player. Swink [10] explains this implicit loop for moment-to-moment play as a correction cycle that demands emphasis on game feel for a continuity of perception. An impairment that creates delay or incapability on any of these stages (impairment of the player or the game) would impact the quality of experience.

Game accessibility requests compassion from the system to bridge the gap between the player and the game in order to reduce the impact of impairment. For example, similar to how a poor visual design of an interface would reduce the usability significantly, a poor visual fidelity of a game world would also reduce the quality of game experience. However, contextual content of games requires far more depth compared to a user interface. The player perceptions of the game world not only stem from the narrative elements that are telling a story of the game world but also the familiarity of the player with the game and game world. Therefore, the player needs a high processing power in order to evaluate all the information they could gather from the game in real time while playing the game. Thus, the system needs to show compassion when player's impairment is getting in the way of their player activity. Without a model that helps breaking down areas that demand processing power and areas to hide delay, it is uneasy to contemplate on how this need could be resolved without frustration.

According to Swink, any delay that breaks the continuity of the experience creates poor game feel [10]. Therefore, the game needs to be responsive for the player inputs. Similarly, any player action in response to the events presented by the game needs to be timely; otherwise, deemed unsuccessful by the game. Hence, the game acknowledges successful behaviour and rewards it while also clearly communicating the consequences upon failure. From a purely game design point of view, this makes sense because facing consequences help bringing meaning to the choices. However, from a player-centric design point of view, especially for players with motor and/or cognitive impairments as in PD, how much time should be evaluated as the time-frame for "timely response" is unclear. The procedures of the game should be forgiving with a suitable error margin and compensating for delays as motor-cognitive processes in player's mind may take longer than an ordinary player.

Research shows that quality of life for PD patients drops over time meaning that activities of daily living, such as dressing, grooming, bathing, self-feeding and functional mobility are jeopardised as disease develops [23]. Even at earlier stages various disturbances and impairments limit the ability of the users while performing tasks that are considered simple; rendering many games inaccessible for this player base. The disturbances and impairments that are commonly observed across PD patients are sensory sensitivities, motor impairments, cognitive impairments and emotional sensitivities. Many of these impairments, especially in early stages of PD, show close similarities to age related changes that are commonly observed among older adults or other health related situations, such as stroke patients [14].

B. Impairments due to Parkinson Disease

1) Sensory Sensitivities: Sensory difficulties include not only hearing or vision problems as mostly seen with elders [18], but also sensitivities for sensory overload due to visual and sound stimuli. Occupational therapy for PD advise reducing visual stimuli by reducing confusing patterns (striped-checker), strong colours and hues, and simplifying the load by preventing contrasting visuals and clutter [22]. Visio-spatial disturbances and strong contrasts cause freezing while clutter overloads cognitive processes with a need of strategising and replanning. There are no specific sensitivities reported about audio; however, the use of metronome and inducing percievable rhythm into daily life are presented to be useful to enhance motor abilities [22].

2) Motor Impairments: Main motor impairments observed in PD are trembling fingers and hands (tremor), rigidity, slowness in movement (bradykinesia), and gait problems [22]. Subtle slowness in movement, postural change and gait problems are also seen in elders even though the scale of these differ from PD. Trembling fingers and hands, especially depending on the scale of movement can make it very hard to use an input device or perform button presses while the slowness in movement can increase the response time.

3) Cognitive Impairments: Cognitive impairments that are commonly seen with PD are learning and retaining information (working memory), concentration and attention, and executive functions. Executive functions are a set of inter-related cognitive processes that are essential for goal-directed behaviours [1]. Even though they are heavily related to cognitive domains, motor skills and connection between cognitive and motor skills are the main reason why they are absolutely necessary for activities of daily living [1][23]. In order to preserve gait, a person needs to evaluate their surroundings, strategically decide a path of movement, shift their weight and meanwhile check their balance. If they come across an obstacle, they should be able to stop executing their plan and rework a new plan similar to the correction cycle mentioned before. This means all six executive functions are actively used during a simple walking task: attention, inhibition, planning, reasoning, shifting (flexibility), and working memory. Gait disorders share similar issues originating from deficiencies in executive functions and also observed among older adults. This means impairments in executive functions also develop among older adults, perhaps milder than PD. In addition to a previously identified need for task creation frameworks to facilitate purposeful use of games for special demographics [4], it has become appearant that there is a need for a design paradigm that draws attention to the abilities and limitations of the players. Therefore, the player-centric design model is developed to support and inform designing games for rehabilitative and preventative therapies for PD.

III. PLAYER-CENTRIC DESIGN PARADIGM

Game design is perceived to be player centered; however, authorial intent and its dissonance with the nature of designer's perception of the player may jeopardize this. This section discusses player-centric design while suggesting a player-centric model for designers to explore and understand the human nature of their target audience.

A. Player-centric Design

It is commonly thought that player-centric design is an extension of the user-centred design. On one hand, one can argue that game design is inherently player-centric since it always questions what the player is doing, what they are allowed to do, objectives, rewards, consequences for the player, and how the player is supposed to feel during the gameplay experience. Therefore, it would be unfair to suggest that game design is not player-centric. On the other hand, game design practice is a creative endeavour as well, meaning that it could carry a separate authorial intent due to its creative nature. Therefore, the intent of the designer in making of a game may be slightly different than the intent or expectations of a player while stepping into the magic circle of the game. Besides, game design practice could also be perceived as play-centric, putting emphasis on gameplay over other elements of the game. A designer can argue for a different act between challenge and motivation, intended difficulty of the game or intended interaction for an intended experience based on the authorial creative endeavour or play focus. In response to all these arguments, player-centric design is centred with empathy to the player and aims to provide a positive experience to the player despite contradictions with creative pursuit [9]. Therefore, player-centric design puts the player before the creator.

B. A New Model for Player-centric Design

The model seen in Fig. 1 attempts to merge player's perspective with formal elements of games in order to enable a deeper discussion on game design challenges for special demographics. Literature has a few well recognised frameworks for game design—Schell's, Fullerton's and Adams' [9][19][20]; however, these do not allow to closely look at underlying concepts of games or dismantle a complete game experience into its thinner slices. Such granularity would be very helpful for designing games for impaired players.

Schell's design tetrad remains quite high level while Fullerton's is too low level without any visible interplay between and above the formal elements or much consideration on player's stance [19][20]. On the other hand, Adams' model is based on interaction design, presenting an interaction model between the player and the game, yet does not encourage explorations on sensory complexity of games more than perceiving them as user interfaces [9]. Although Adams' could be a much clearer approach that allows the use of existing HCI research for games, it creates further ambiguity on the potential needs of games on a visual and contextual level. Gerling et. al's "extended model" presents a useful basis for designing for older adults [18]; however, the analysis lacks a wider perspective on the interplay between the elements of the model as a game design paradigm and a player-centric focus for the designer to explore the position of the player in this context especially for the needs/issues of special demographics.

In order to create a player-centric focus, the proposed model (Fig. 1) combines layers of player experience—as inspired from Garret on the planes of user experience [7][8]—with Adam's game design model [9]. The layers of the model and elements of these layers are also discussed further in the following subsections.

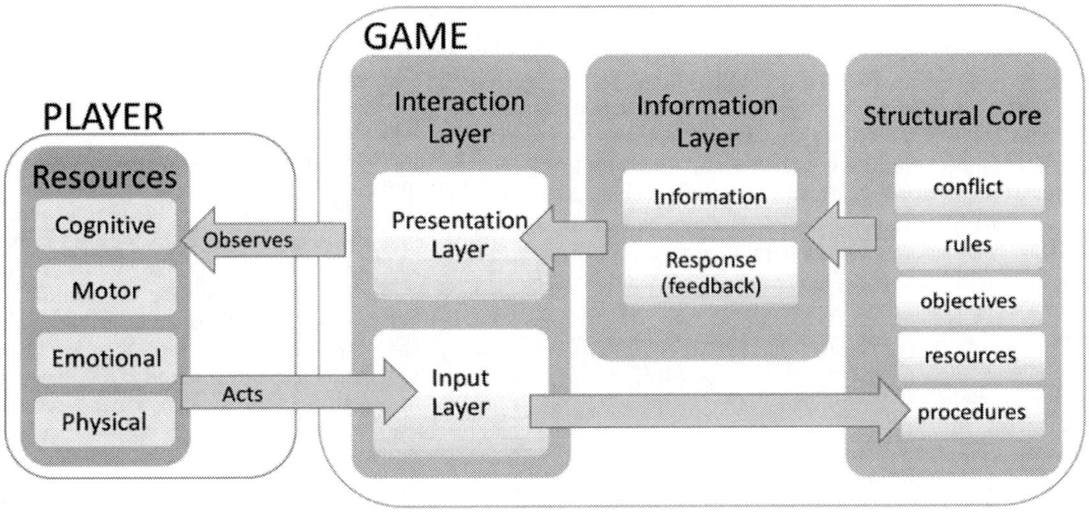

Figure 1. A player-centric game design model favouring the capabilities of the player and perceiving the system from player's point of view

1) Interaction Layer: Interaction layer represents the overlapping space of game world and player's world. Via this layer, player sends input to the game world, observes the results of their input and receives a response from the game world. Interaction takes place between the player and the game world. It is either started by the player via an input, or by the game world via an event presented with the presentation layer. When it is initiated by the player, the game responds; when it is initiated by the game via presentation layer, the player responds via input layer. In both cases, information layer feeds the presentation layer with necessary data. Interaction layer is composed of presentation layer and input layer.

a) Presentation Layer: Presentation layer can also be thought as sensory layer encompassing audio, visual and haptic presentation of the current status of the game. This includes continuous presentation of the game world and the game's response to the player inputs. Any feedback generated in response to the player inputs is presented by the presentation layer. Clarity of presentation and a suitable composition of audio-visual elements are essential for readability. Based on movement advice of PD handbook [22], a suggestion to facilitate performance of motor skills is using sensory stimuli that provides a perception of rhythm. Since the cognitive difficulties include attention and concentration, any support to conciously activate attention and maintain concentration is very useful in daily life, therefore for the presentation layer as well.

b) Input Layer: Input layer is responsible for the interaction device, input techniques, clarity of input mapping, directness, sensitivity and consistency of input. A common goal for a game controller is effortless use with which the input device feels like the extension of the body [10]. It is important to note that PD has some symptoms, such as tremor, bradykinesia or hypokinesia that may cause difficulty in using an input device or perform an input action within a required time-frame. Therefore, additional research into input devices and interaction modalities would be useful. Moreover, further research on familiarity of the player group, mental model, and restrictions of disease stages for input modality is expected to improve the design.

2) Information Layer: Information layer sits in the middle of presentation layer and structural core of the game, and interprets outputs of the core system in a readable format for the player. Cues for meaning making (semantics), affordances and limitations for the player [12], contextual visual material, data organisation, response of the game (feedback for the player), and any information, such as score, status, outcomes, etc., belongs to this layer. Even though the information is generated by the core layer of the game, its interpretation is handled in information layer and passed to the presentation layer for the player to see.

3) Structural Core: Structural core of a game comprises formal elements [19]. Both the information layer and the presentation layer are dependent on the core structure of the game while also creating meaning for it. From player's point of view, the core of the game may be completely invisible as their perception is shaped by the presented information (based on how it is interpreted by the information layer). Therefore, discoverability, learnability and consistency of the system need to be resolved at this layer so that relevant information could be fed to the information layer.

Procedures are the first point of interaction with the input from the player. They are integral for moment-to-moment gameplay and define the chain of moves to perform actions in the game. Reiterating the previous discussion on correction cycle (see Section II, *A* and *B.3)*, a delay in any stage of player performance will make the time-frame of user input longer. For example, double jump could require hitting the jump button twice within a second in order to perform double jump. This seemingly simple action could pose a problem for a person with rigidity or sloweness of movement, who might find hitting the same button twice hard to repeat within a second's window. Another example is the number of steps necessary to do something, such as the steps to be performed to bake a cake. For a person with memory issues, remembering those would be really hard, therefore frustrating to perform. Thus, procedures should be catered for a longer grace time, simple recovery (this does not mean game needs to be easy), shorter and less complex chain of actions for learnability and retaining information. On top of these, additional compassion for impairments would make a big difference. When done right, the core game demands less player resources or compensates for those when necessary. The player resources are explained below in item 4.

Contextual content works with the formal elements and supports meaning making. Objectives, rules, procedures, affordances and limitations become meaningful with the help of contextual content. While being important for engaging the audience, contextual content is also important for maintaining attention and motivation. In order to develop content based on the interests of the target audience (elders, kids, young adults, etc.), a participatory approach or persona studies would be preferable.

4) Resources: Rather than the resources discussed for formal elements of the game, the resources referred in this model is intrinsic player resources. Intrinsic player resources are cognitive skills, motor skills, physical abilities and emotional abilities (physical and emotional stamina). These are within the power of the player and do not belong to the in-game economy. They are not generated by the game or in the game; however, they are brought in and used by the player, yet consumed by the game. They are limited, and refresh time for these vary from person to person. For example, after long hours of play, the physical stamina of a player may drop, and they may not be able to function as prompt as they were at the beginning of the play. This is an example of player spending their physical ability resource. For a suitable design tragetting players with health related difficulties and in order to prevent frustration, designers need to study how these resources are effected by the disease.

C. Strengths and Limitations of the Model

The player-centric design paradigm, as seen in Fig. 1, is developed as a bi-product of design research for game based rehabilitation of PD. Previous sub-sections presented further details on the layers of the model with a focus on constraints for PD. The purpose of the model is to promote further discussion on the elements of game design with a player-centric focus; therefore, the main strength of the work originates from the incorporation of user experience model to ensure this. It is by no means a complete model, yet welcomes exploration in those layers, and encourages analytical thinking towards player-centric design. The main limitation of the paradigm is its theoretical nature even though it has emerged from the design process of a game for players with PD.

The model stands as player-centric, yet the principles of user-centered design and their alignment with the model has not been completely examined. A useful improvement would be development of a set of questions for each layer to prompt the designer while using this paradigm in their design practice. In addition, the discussion on information layer is less explored compared to the other layers and could benefit from further contemplations on the relation between contextual elements and information layer, especially for the potential impacts of these on players' resources. Finally, an additional angle with playability concept and a discussion on how playability relates to this model would be informative.

IV. CONCLUSION AND FUTURE WORK

In this paper, a newly developed player-centric design paradigm is presented to improve game design practice for special demographics. The model has extended existing game design models with inspiration from user experience field in order to present a player focused practice for game design. The layers of the model are discussed by employing a PD focused impairment analysis. Future work includes verification of the presented paradigm with the analysis of existing games and the development of new ones.

REFERENCES

[1] A. Diamond, "Executive functions," *Annual review of psychology*, vol. 64, pp. 135-168, 2013.

[2] D. Levac, S. Glegg, H. Colquhoun, P. Miller, F. Noubary, "Virtual reality and active videogame-based practice, learning needs, and preferences: A cross-Canada survey of physical therapists and occupational therapists," *Games for health journal* 6, no. 4, pp. 217-228, 2017.

[3] A. E. Staiano, and R. Flynn, "Therapeutic uses of active videogames: a systematic review," *Games for health journal*, vol. 3.6, pp. 351-365, 2014.

[4] A. T. Bayrak, B. Wünsche, C. Lutteroth, "Reviewing the Evidence: n Pursuit of a Framework for Parkinson Disease Rehabilitation with Games," *The Promise of New Technologies in an Age of New Health Challenges*, p. 9, 2016.

[5] B. Bonnechère, *Serious Games in Physical Rehabilitation*, Springer, 2018.

[6] J. Nielsen, "Nielsen norman group," [Online]. Available from *https://www. nngroup. com*, 2016.

[7] C. Macklin, and J. Sharp. *Games, Design and Play: A detailed approach to iterative game design*, Addison-Wesley Professional, 2016.

[8] J. J. Garrett, Elements of user experience, the: user-centered design for the web and beyond, Pearson Education, 2010.

[9] E. Adams, *Fundamentals of game design*. Pearson Education, 2014.

[10] S. Swink, Game feel: a game designer's guide to virtual sensation. CRC Press, 2008.

[11] O. Assad et al., "Motion-based games for Parkinson's disease patients," In *International Conference on Entertainment Computing*, pp. 47-58, Springer, Berlin, Heidelberg, 2011.

[12] D. A. Norman, Emotional design: Why we love (or hate) everyday things, Basic Civitas Books, 2004.

[13] K. Salen Tekinbaş, and Eric Zimmerman, *Rules of play: Game design fundamentals*, MIT press, 2004.

[14] M. Balaam et al., "Motivating mobility: designing for lived motivation in stroke rehabilitation," In *Proceedings of the SIGCHI Conference on Human Factors in Computing Systems*, ACM, pp. 3073-3082, 2011.

[15] A. C. de Barros, J. Cevada, À. Bayés, S. Alcaine, B. Mestre, "User-centred design of a mobile self-management solution for Parkinson's disease," In *Proceedings of the 12th international conference on mobile and ubiquitous multimedia*, ACM, p. 23, 2013.

[16] L. Caroux, K. Isbister, L. Le Bigot, N. Vibert, "Player–video game interaction: A systematic review of current concepts," *Computers in Human Behavior* 48, pp. 366-381, 2015.

[17] K. Isbister, and F. "Floyd Mueller. "Guidelines for the design of movement-based games and their relevance to HCI," *Human–Computer Interaction* 30.3-4, pp. 366-399, 2015.

[18] K. M. Gerling, F. P. Schulte, J. Smeddinck, M. Masuch, "Game design for older adults: effects of age-related changes on structural elements of digital games," In *International Conference on Entertainment Computing*, Springer, Berlin, Heidelberg, pp. 235-242, 2012.

[19] T. Fullerton, Game design workshop: a playcentric approach to creating innovative games, CRC press, 2008.

[20] J. Schell, *The Art of Game Design: A book of lenses*. AK Peters/CRC Press, 2014.

[21] A. Kultima, "Game design research," Proceedings of the 19th International Academic Mindtrek Conference, ACM, 2015.

[22] A. Aragon, B. Ramaswamy, J. C. Ferguson, *The professional's guide to Parkinson's disease*, Parkinson's Disease Society, 2007.

[23] G-M. Hariz, L. Forsgren, "Activities of daily living and quality of life in persons with newly diagnosed Parkinson's disease according to subtype of disease, and in comparison to healthy controls," *Acta Neurologica Scandinavica* 123.1, pp. 20-27, 2011.